Praise for
The Balance Within
———◀◯▶———

"*The Balance Within* is a tour de force of the past, present, and future of our knowledge of mind/body interactions and stress. Dr. Sternberg, a leading expert on the interaction of the endocrine and immune systems, writes beguilingly. A knowledgeable and entertaining tour guide, she makes complex issues clear. Dr. Sternberg takes us from the origins of medicine in Greece, to early medical schools in Padua, to modern research in Montreal and the U.S. She lucidly describes how we came to appreciate the physiology of stress, how the mind influences the body, and how the body affects the mind. More than food for thought, this book is nourishment for those curious about mind and body."

> — David Spiegel, M. D., Professor and Associate Chair of Psychiatry
> and Behavioral Sciences, Stanford University School of Medicine

"The author has undertaken the daunting task of bridging a chasm between what we know as the scientific basis behind disease and what we don't know about how our brains can influence this science to make our physical and mental health either better or worse. She takes us on a fascinating trip, describing difficult scientific concepts in easily understood terms and liberally uses colorful analogies to bring the science into a reality we can all appreciate. However, the journey is not yet at an end, in that knowledge is continuously being added, filling in gaps presently occupied by guesses. One can only hope that Dr. Sternberg will continue to write, acting as our tour guide to interpret the science and connect it with our daily lives in such a way that we can learn to seek and find help within our being."

> — Frances K. Conley, author of *Walking Out on the Boys*

"Dr. Sternberg's book is a dazzling tour of a most promising area of neuroscience–the interface between the immune system and the nervous system. The pathways by which the nervous system controls stress responses, inflammation, and other hormonal and immune system events, and the feedback mechanisms by which immune system chemical transmitters regulate the nervous system, have only recently been discovered. This area of research, in which Sternberg has been one of the world's leading scientists for at least a decade, is leading to new understandings and treatments of the stress-related diseases of modern life, including chronic fatigue syndrome and fibromyalgia."

> — Elliot S. Gershon, M. D., Professor of Psychiatry,
> The University of Chicago

The Balance Within

The Balance Within

THE SCIENCE CONNECTING
HEALTH AND EMOTIONS

Esther M. Sternberg, M.D.

W. H. Freeman and Company
New York

TEXT DESIGNER: Diana Blume

Library of Congress Cataloging-in-Publication Data

Sternberg, Esther M.
    The balance within : the science connecting health and
emotions / by Esther M. Sternberg.
       p. cm.
    Includes bibliographical references and index.
    ISBN 0-7167-3479-6
    1. Medicine, Psychosomatic. 2. Psychoneuroimmunology. I. Title.

    RC49.S655 2000
    616'.001'9–dc21   99-086775

Printed in the United States of America

First printing 2000

For my parents,
Joseph Sternberg, M.D., and Ghitta Sternberg,
in memory

Contents

Preface

*T*he idea for this book grew out of a question. Why was it, I wondered, that those of us who study brain–immune connections have seemed ostracized from the rest of the scientific community? Although our research findings clearly show that the brain and the immune system communicate, researchers in the hard sciences not only ignored but also outrightly rejected our field. Why was that? I began to discuss this question with historians of science and realized that our experience was not unique—in fact, it was rather commonplace. Whenever a new field comes into being, it comes up against the older dogmas. So the resistance that we felt was real and steeped in traditions going as far back as Galileo, Copernicus, and beyond. This insight helped me decide to take a proactive stand, to step outside the immediate bounds of the laboratory and systematically try to bring the findings of this discipline out of the fringes and into the mainstream of scientific consciousness. Not surprisingly, this turned out to be a more daunting task than I had imagined. Whenever one tries to change prevailing opinion, resistance is inevitable. Fortunately, I was supported by a few other like-minded spirits and bolstered by the burgeoning wealth of scientific findings pouring out from more and more laboratories at the cutting edge. For, ultimately, what convinces scientists of the truth is an accumulation of hard, irrefutable facts, collected from rigorously performed experiments. These sorts of studies were—and are—exponentially increasing in the field.

Several events fortuitously occurred that helped me to showcase the field and explore the question in a more systematic way. In 1995, Dr. Samuel "Don" McCann, neuroendocrinologist and champion of neuroimmunomodulation (as the field of mind–body communication was beginning to be called), asked me to help host an international meeting of the Society of Neuroimmunomodulation at the

National Institutes of Health. In conjunction with organizing the scientific side of that meeting, I needed an event to open the conference. A serendipitous snafu then led to the next step. An invitation destined for psychologist Dr. Sheldon Cohen of Carnegie Mellon University in Pittsburgh to speak about his studies on stress and the common cold erroneously went to Dr. Sheldon Cohen, Emeritus Scientist at the National Institute of Allergy and Infectious Diseases and now historian of science at the National Library of Medicine. Allergist–immunologist Cohen declined the invitation but was intrigued by the subject and suggested that I put on an exhibition at the National Library of Medicine (NLM) on brain–immune connections and disease. He put me in touch with Dr. Elizabeth Fee, just arrived as the new Head of the History of Medicine Division of the NLM. Liz responded to the project with enormous enthusiasm, and the 1996–1997 critically acclaimed NLM exhibition "Emotions and Disease" was the result. Although many were involved in putting this project together, I am especially grateful to Dr. Donald Lindberg, Director of the National Library of Medicine, who took a risk and backed the exhibition to its successful completion. In addition, I thank his wife, Mary Lindberg, whose heartfelt positive response to the finished product meant a great deal to me. Thanks, also, to the vision of the exhibition sponsors, David Mahoney of the Charles A. Dana Foundation, Lynn Gordon of the Fetzer Institute, and Dr. Robert Rose of the John D. and Catherine T. MacArthur Foundation.

Working with the historians involved in the exhibition was like taking a crash course in history, the social sciences, and the anthropology of science. Liz Fee has remained my counterpoint in the two cultures, constantly reminding me, with mischievous delight, that the scientific view is not the only one, that inexorable progress toward a golden technological goal is not necessarily the rest of the world's view of perfection. I learned a great deal, too, from Dr. Ann Harrington, Professor of the History of Science at Harvard University, and Dr. Ted Brown, Professor of the History of Medicine at the University of Rochester, both co-curators of the exhibition. Historians Drs. Declan Murphy and Molly Pyle gave me their historical perspective and encouraged me during readings of the earliest versions of the manuscript of this book. And I am grateful also to Dr. Vyacheslav

(Koma) Ivanov for our many conversations and for his telling me the story of the Hittite king.

These interactions helped me to see my field in the context of the culture in which we live and the deep historical roots from which it grew. This is not a viewpoint that most scientists are taught. We tend to do science as if it springs from the test tube, in a vacuum, untouched by its cultural surroundings. Historical time in science is often measured in weeks or months—certainly not decades. Something that is more than a year or two old is long out of date and often considered worthless; areas of intellectual endeavor not involving technology may be viewed as soft. Many scientists are thus oblivious to the societal and historical context that shapes our thinking and our choice of problems to be solved. My discussions with all the historians with whom I worked have helped me to shed my own scientific bias, thus laying the groundwork for my writing this book.

Several other fortuitous events also occurred around that time, preparing me for the book. Among the most important influences was my membership in the John D. and Catherine T. MacArthur Foundation Mind–Body Network—a group of scientists assembled by Dr. Robert Rose, the Director for Mental Health Policy and Research at the Foundation. One clear criterion for participation in this network was an ability to let go of the boundaries of one's own discipline and a willingness to learn and respect the framework of another's field. It was participation in this network as much as anything else that broadened my thinking beyond a narrow scientific focus to one encompassing psychology, the social sciences, and history. It has been a privilege to interact intellectually and scientifically with the brilliant and articulate members of this group, whose expertise spans social and physiological psychology, neuroimaging, sleep, endocrinology, immunology, and the history of science. Our many hours of lively discussions and grappling with research problems affirmed to me that complex problems in science and medicine, such as the mind–body connection, must be studied from all sides, from top-down and bottom-up, and by applying the tools of different disciplines to piece together coherent answers. In this context, I am especially grateful to Dr. John Cacioppo, a fellow member of the Mind–Body Network, for his invaluable help in reading the manuscript.

Another network member, Dr. J. Allan Hobson, Professor of Psychiatry at the Harvard School of Medicine, specifically encouraged me to begin writing this book when it was just a glimmer of an idea. Allan also stimulated me to expand my style beyond the scientific, teaching me that, in writing, one should "let a thousand flowers bloom." Eventually, it was Allan who put me in touch with my first editor at W. H. Freeman and Company.

Around this time, I was writing an article for a special issue of *Scientific American* on the science of mind–body connections. With my co-author, Dr. Philip Gold, I had struggled to summarize the field in a few densely written pages, aimed at the magazine's traditional audience. Since W. H. Freeman is affiliated with *Scientific American*, both this magazine article and Allan Hobson's recommendation happened to arrive on Jonathan Cobb's desk simultaneously. Jonathan, then Senior Editor at Freeman and editor of the Scientific American Library series, called me one day from New York to say he was going to be in Washington and wanted to meet with me about doing a book. It was every writer's dream!

In the course of constructing the proposal over the next several months, it was Jonathan who helped me find my voice. He worked tirelessly through phone and fax and many long, late-night conversations first to shape this book and then to draw out my writing style. Without him, this book would not have been born. Unfortunately, Jonathan left W. H. Freeman before the project was even signed or sealed, let alone delivered. However, I was lucky enough to find continued guidance and encouragement from all the staff at Freeman, especially my editor, John Michel, and the President of Freeman, Elizabeth Widdicombe. I am especially grateful to Liz for her heartfelt support and personal interest in the book from the very start. I am deeply indebted to John for his painstaking work in shepherding my thoughts and sometimes long and winding phrases. John pruned this manuscript to its current concise and harmonious form. He spent many long, patient hours in conversation to calm my anxieties, and if he felt the need to "shorten," whether a word or a whole segment, he did it gently, always balancing his excisions with an encouraging comment.

Besides these intellectual and professional events, there were larger life events during this period that, more than anything else, were the driving emotional forces behind this book. These were my mother's death and, before that, my father's death. My mother was diagnosed with breast cancer less than six months after my father died from a long illness. Her health deteriorated in the year before I wrote the book, the year that I was developing the "Emotions and Disease" exhibition. And so, I learned to see the science and medicine I practiced professionally through the eyes of a patient and family member. I was forced to face the awful feelings of helplessness of being the family member of a dying cancer patient. Alongside my mother, I experienced the frustrations of high-tech medical care. Even as I struggled to convince her of the benefits of modern science, I could understand her desire to throw it all away, to take charge of her health, and to search for strength in her own form of spirituality. Begun at a time of mourning, this book, then, helped me to explore the processes of my own grieving, and it gave me solace.

However, the book grew not only from my parents' deaths but also from their lives. The gifts my father and mother left to me were the way they lived and saw the world: my father the scientist and physician; my mother the artist and, in later years, the writer. My father taught me always to question, confirm, and analyze. My mother taught me "to see a world in a grain of sand," and to "view every sunset as if it were my last." They both taught me to rejoice in life, to notice and appreciate every flower and to delight in mountain views and sunsets.

When I began this book, I started from the perch of the scientific skeptic. But as I wrote and thought and talked to others, I began to understand that the viewpoint of modern science carries with it a kind of arrogance that has in many ways rightfully alienated the popular culture. The notion that just because we don't understand how something works, it can't be real, is too pervasive in our modern way of practicing science and medicine. But as much as it is a recognition of the popular belief in the power that mind has over body, this book is a tribute to the scientists, the physicians, and their discoveries that have helped us reach the point at which we are today. Even the greatest skeptic must now admit that a wealth of evidence exists to prove in the most stringent scientific terms that the functions of the mind do influence the health of the body and that sickness in the body can affect our moods and emotions through molecules and nerve pathways.

This level of proof of the myriad connections between the brain and the immune system was needed before the two cultures—the popular and the scientific—could begin to respect and talk to each other.

There are many pioneers in the field of brain–immune communication, and it would have been impossible to name all the scientists and great discoveries that have contributed to our current state of knowledge. They all worked against great odds, especially in the early days when researchers in the traditional sciences scoffed at this area. Without these pioneers, there would be no story to tell. These brave souls included Drs. Bob Ader, Nick Cohen, George Solomon, and Herbert Weiner, who all worked closely with Norman Cousins— an inspiring and charismatic voice for the field of psychoneuroimmunology from its start. On the neuroendocrine side, Drs. Don McCann, Novera Herbert Spector, Jim Lipton, and Seymour Reichlin worked to convince their immunologist and endocrinologist colleagues of the validity of this science. Finally, I cannot refer to the beginnings of this field without a tribute also to Dr. Candace Pert, who accepted me into her lab when I first arrived at the National Institutes of Health (NIH). Candace, a true pioneer in the field, was one of the first neuroscientists to recognize that the brain and the immune system might communicate at a molecular level. I am indebted, too, to my colleagues at NIH, Drs. Ronald Wilder, George Chrousos, Philip Gold, and Steven Hyman.

In the pages that follow, I have only superficially touched on, or omitted altogether, many areas of health and disease that are impacted by this mind–body science. To do justice to all these areas and the researchers who did so much to uncover these connections would require volumes. Thus, I have only briefly described the important work that has been done on mind–body connections in AIDS—work by researchers such as Drs. Neil Schneiderman, Margaret Chesney, Margaret Kemeny, Shelley Taylor, and John Fahey. I have only briefly described investigations into the role of mind–body connections in cancer by researchers such as Drs. Fawzy Fawzy and David Spiegel. I have omitted many important details of large areas of research, such as that on the relationship among stress, steroid hormones and tuberculosis; thymus–brain connections; the effects of prolactin on immunity; the expression of cytokines in the brain and neural–glial interactions; multiple sclerosis; the role of cytokines in fever; melatonin–pineal gland–immune connections; and sleep and nutrition. And I could not mention by name all the members of the next generation of researchers who are taking this field to its next level of molecular and neuroanatomical precision and clinical applications. In a way, this long list of omissions is a testament to how large and varied this field has become since its modern reincarnation only a few decades ago. For those interested in exploring such areas in greater depth, I have included, at the end of the book, a bibliography of suggested readings.

In gathering material for this book, I talked to many scientists and physicians who shared with me their personal memories of places, events, and people I describe. I am grateful to Dr. Soreen Sonea, Professor Emeritus and former Chairman of Microbiology and Immunology at the University of Montreal. Dr. Sonea was a dear friend and colleague of my father and colleague of Hans Selye. His memories and our long conversations helped me piece together a more personal remembrance of the working environment in which Selye developed his famous theories on stress. I recently also had the opportunity to meet two of Dr. Selye's former students, Dr. Arpad Somogyi, now Head of the European Commission's Unit on Evaluation of Health Risks, and Dr. Istvan "Steve" Berczi, Professor, University of Manitoba, whose delightful tales also helped to confirm my memories. I am grateful to Dr. Berczi for providing the wonderful photograph of Dr. Selye from his collection. I also thank Drs. Douglas Arnold, Jack Antel, and Mark Angle for taking me on a personal tour of Wilder Penfield's operating theater in the Montreal Neurological Institute (MNI), virtually unchanged from the way I remembered it when I first saw it as a medical student. I am grateful also to Dr. Bill Feindel, former Director of the MNI, for his historical insight from my student days. I am indebted to Drs. Giulia Perini and Giovanni Silvano for arranging a tour of the anatomical dissecting theater in Padua at the Palazzo del Bo and for providing me with the beautiful books and insights into the history of the University of Padua. I am also indebted to the curators of the University of Padua Orto Botanico and Centro per la Storia for providing me with the exquisite prints of the botanical gardens and the anatomical dissecting theater of their great university. I am grateful to the National Academy of Sciences, Washington, D. C., for allowing me to photograph the Academy's mural of

Prometheus. Thanks, too, to Bill Mapes for providing the drawings of the brain–immune connection. I thank Dr. Dirk Hellhammer for his tours and stories of Trier, and Drs. Noel Rose, George Wick, and Konrad Schauenstein for their remembrances of their seminal studies on chickens. I appreciate also the personal remembrances of Drs. Robert Dantzer, Steve Maier, Linda Watkins, and Dwight Nance on their research on sickness behavior and Dr. Bob Rose's personal insights into his studies on job stress among air traffic controllers. And I am grateful to Eleanor Cousins for her conversations with me and her remembrances of her husband, the late Norman Cousins.

Serendipity often plays a role in scientific discovery, and it certainly played a role in the cover art for this book. Had I not walked into a local copy center at the moment the image of Larry Kirkland's stunning installation was being photocopied, the book would not have had the cover (or even the title) it does. Both title and image perfectly sum up the historical, scientific, and philosophical themes of the book. I owe a great debt of gratitude, then, to the artist Larry Kirkland, as well as to J. Stark Thompson, the CEO of Life Technologies, where this installation hangs, and to photographer Robert Lautman for permission to use this work. Many thanks also to Sloane Lederer, Director of Trade Sales and Marketing at W. H. Freeman, for coming up with the perfect title.

In all the phases of the writing of this book, I relied a great deal on a wide and warm circle of family and friends who stood by me, encouraged and cajoled me, and loyally listened to my readings of the manuscript. My daughter, Penny Herscovitch's, genuine enthusiasm seasoned with honest criticism helped me to remember never to let my writing slip into "cheesiness." My sister, Aline Petzold, has always been there for me at every turn and has lived through the process of writing this book as if it were her own. Drs. Anne and Morris Wechsler, and Becca, Morty, and Randy Goldsmith have also lived this book at every stage and devotedly listened to many versions of the manuscript with support and enthusiasm. I am grateful to my dear friends and colleagues, Drs. Jan-Ake Gustafsson and Samuel Page, for believing in me and encouraging me throughout this project. I appreciate, too, the encouragement from my friends, colleagues, and editors in other capacities, Drs. Orla Smith and Alan Hammond. I thank Diana Lady Dougan for our many conversations and for her special input on the kind of information on health and the mind–body connections that a nonscientist audience might seek. And I thank Dr. Peter Watson for his unswerving advice and encouragement. My friend and one-time neighbor, Susan M. Forward, has been my invaluable sounding board for many of the ideas in this book. She has been my window onto the popular culture's view of health and the many ways that believing can make you well. My many long conversations with Susan, her enthusiastic response to the earliest versions of the manuscript, and her capacity to truly listen and ask probing questions helped me frame many of the health-related issues in the book. Her honest criticism reminded me when too much of my scientific side took over.

Finally, I am grateful to my friends Tarja and Dean Pappavasiliou for stopping by on that rainy March afternoon just after I moved into the house next door. I thank them for the genuine and warm European hospitality that led them immediately to invite me to spend a vacation with their family in their cottage in Crete. It was that idyllic experience of the sun and sea in the village of Lentas, combing the archaeological ruins of the temple to Aesclepius and sipping coffee at the cafes on the beach, that helped me to heal and became the inspiration for the theme of this book.

CHAPTER · 1 ·

Emotions and Disease

Molecules and Ancient Myths

—————◀◯▶—————

*N*estled at the top of a brown stony hill above the modern Cretan village of Lentas, at the intermingling of cool sage mountain air and warm salt sea breezes, are the ruins of an ancient temple to Asclepius, the Greek god of healing. The temple's two remaining pillars stand like sentinels marking north and south, forming the narrow end of a once oblong colonnade, built in exact alignment with the sun's path. It is a few meters above what was once the source of a natural spring; ancient priests used these waters, and prayer, music, sleep, and dreams to cure the sick. The warm white marble columns still reflect the early morning sun against the blue Mediterranean below. On a flat terrace, just beyond the temple, a mythical animal with the head of a horse and body of a fish, patterned out of smooth, round, white and black pebbles, hides the floor of the priests' coffers of this once rich sanctuary. And the village people, who still live as one with the rhythms of the sea and sun, know, as their ancestors knew, that emotions and health are one.

As the wind and sun eroded that first ancient shrine, and dried its healing source, something also happened to the world beyond the village. Our faith in the healing power of the spirit also waned; and the god of science and medicine became a much harder, more impersonal god than the fatherly Asclepius. When did we modern scientists and physicians lose the knowledge that was so much a part of these ancient teachings of medicine? And why has the road back to acceptance of this wholeness taken so many centuries to travel?

◄○►

The temple and the ancient town below, once a busy crossroads port called Leban, first stopping point between Egypt and Greece, flourished around 400 B.C. This was about the time that the great Greek physician Hippocrates, whose oath still underlies the principles of modern medicine, taught that health lay in a balance. And Asclepius, the Greek god of healing, with his daughters, Hygeia and Panacea, symbolized all that was essential in this balance—healthy diet, pure waters, exercise, and support of friends and family. The concept has survived to this day: "hygiene," or cleanliness, is still the first step in preventive medicine, and "panacea" still means heal-all. But essential, too, were the emotions, as well as soothing activities that calmed them—sleep, music, and prayer. So integral to the healing of the body was the mind that the god of medicine carried a staff with the symbols of both intertwined: Asclepius carried in his left hand the caduceus, a wooden staff with a serpent curled around it, an ancient symbol of body and soul, and today the universally recognized symbol of medicine.

Facing east, with your back to the pillars of the ancient temple, squinting into the already hot morning sun, you can see a tiny stone church just 100 meters away. Amidst the scrub and stones and fallen pillars, a lone gardener lovingly tends the shrine. To reach it, scramble over the loose rocky soil, down the steep terraced ridges of the temple, past where the ancient source used to run. Climb up an embankment, through lavender and prickly bramble, and you are there. It sits atop its own little hill, amidst another pile of ruins—also the remains of a place of worship, built later than the temple but before the newer church. Here the gardener pulls the weeds from between the cracks and crevices of the flat square stones paving the earth, stones that must have formed the floor of the Byzantine basilica that stood here more than 1200 years ago.

The newer church, itself nearly 100 years old, is not much larger than a small room and not much taller than a man. Bending low to enter the single doorway, you find the impact of the cool air within to be an immediate relief from the unrelenting sun that beats down

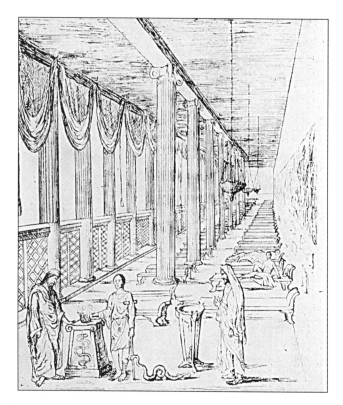

More than two thousand years ago, the Greeks understood intuitively that emotions and health are one. Asclepions like the one above were named after the Greek god of healing, Asclepius, who symbolized all that was essential in this balance—healthy diet, pure water, exercise, and support of friends and family. Essential, too, were the emotions and soothing activities that calmed them—sleep, music, and prayer.

upon the hillside. As your eyes acclimatize to the darkness, you realize that you are surrounded by a surprising array of brightly painted red, gold, silver, black, and brown icons—faces of the Virgin Mary and the baby Jesus. The flicker of votive candles left by the devout villagers lights the pictures, large and small. The villagers come alone to pray or throng here in large numbers whenever a priest passes through the tiny village on his rounds. Old and young alike scramble up the steep, well-worn footpath, bending almost horizontally to grasp at branches, steady their gait, and keep from sliding backward as the stony soil gives way.

But now, you are alone. The gardener has left. Only swallows remain, flitting in and out from shaded eaves under the church's red-tiled barrel roof. Peering through the narrow door toward the shining sea and village far below, you glimpse the sun, almost straight above. It must be close to noon—too hot for any mortal being.

Clambering down the hillside, past riotously bright red, pink, and fuchsia bougainvillea attached to white stucco walls, you reach the village square. The villagers are all asleep. Closer to the sea, the salty and faintly fishy smell mingles with the eucalyptus whose branches provide the first shade over the steepening lane. At the last row of houses, a flight of narrow, uneven stucco steps twist like a pile of children's blocks to the pebbly beach below. Finally, in the narrow spaces between the sharp white walls, is the sea—a blue so intense, so azure, so deep, that it seems unreal.

Here, along the cove, the fishermen live in balance with the elements. The rhythms of their lives are set by the rhythms of the day and of the seasons. They watch for signs of storm, and tide, to know when to tie their boats closer in and tighter to the shore. And they watch the changing seasons, to know when to retreat to a town high up in the mountains behind the village, when the rainy season comes. They watch the stars and know, just as their forbears knew, the names of all the constellations and of our galaxy. For here, by the shore in the dark, dark night, you can see the Milky Way as clearly as did the ancient Greeks, who named this faint spill of white that stains the night sky *galaxias,* from *galactos,* the milk that it resembles.

Today, the villagers have adapted, but only somewhat, to the trickling tourist trade—the few brave souls who have trekked over the daunting mountains that hem this tiny village up against the sea. Each house along the pebbled cove sports a porch with brightly colored tables and chairs, painted blue or red or green. Here villagers and tourists wile away the day sipping thick Greek coffee and lemonade made with water from what is left of the source. They nibble at Greek salads, creamed eggplant, cucumber yogurt "tzatzikis," or grapeleaf-wrapped dolmades, and watch the sea. The old men, long since unable to gain a living from their boats, sit at their favorite tables, playing backgammon, and time and again beat the younger men, stronger than they, but with yet much to learn about the game. There is a richness to the social fabric of this place: children run free, but are watched by a dozen mothers' eyes; gnarled old men and women, though walking with canes, are still able to climb the steep slopes daily to visit neighbors and to pray at the shrine above the town. What happened to our modern world, where isolation has replaced social support, where technology has broken the bond between doctor and patient in healing, where the role of emotions in health and disease has been too often cast aside?

◄○►

Perhaps it was the discoveries of twentieth-century physicists—the ever-smaller particles that make up the physical universe—that turned our thinking away from emotions and toward a more quantifiable physical world. "If you can't measure it, it isn't real" became an unspoken dogma of modern science and by extension of medicine. Perhaps it was the discoveries of the sixteenth-century anatomists—the abnormalities of anatomy underlying many diseases—that made us assume that illness could spring only from concrete and visible distortions of anatomy, not ephemeral and invisible distortions of mind. "If you can't see it, it isn't real" became another tenet of modern medicine. Or was it Copernicus's revolutionary doctrine of an earth revolving around the sun, and not vice versa, that made us all discard the idea that invisible, uncontrollable fate—and by extension, invisible, uncontrollable emotions—held power over our daily lives, our health, and our deaths?

Most historians and philosophers agree that it was the teachings of the seventeenth-century French philosopher René Descartes that ushered in the thinking of the modern age and began the unraveling of the ancient link between emotions and health. In his reaction to the religious wars and the resulting turmoil that spread across Europe for most of his adult life, Descartes formulated the concepts of rationalism and the necessity of visible proof that were to become the founding principles of modern science. In that era, emotions seemed a thing of magic, fleeting and undefinable in the framework of the science of the day. In Descartes's orderly division of the world into rational and irrational—provable and unprovable—emotions and their relationship to health and disease clearly fell into the latter domain. And there they remained until scientific tools powerful enough to challenge the categorization could rescue them.

Whatever the precise combination of philosophies that contributed to this mind-set, the latter-day philosopher-scientists left a legacy so ingrained in their successors' way of doing science that many scientists don't even realize it's there, and they certainly don't recognize where it came from. There has been, for example, an assumption at all levels of science, from basic laboratory to national research funding institutions, that unless a study is focused and narrow, it is bad science. Similarly, if the work does not fit into a clearly defined and narrow discipline, it is bad science. This tide is turning now. But as recently as the 1990s, some leading scientific policy makers and academicians could still be heard to say that stress research had no place in the structure of AIDS research funding, some have even been quoted as calling it "puffery." Such resistance to new ideas has many precedents in science. James Lister and Louis Pasteur were at first excluded from academic honor societies and laughed at for their theories on sterilization, vaccination, and pasteurization. But there was yet another reason for the academic medical community's rejection of the science of the mind–body connection. Beyond its newness is its oldness. That it is embedded in the popular culture made serious scientists shy away from it for many years, for fear they would be branded fakes.

Only very recently has this mind-set begun to thaw—with considerable pressure from the popular culture. The overemphasis on a narrow focus, combined with a fascination with technology to the exclusion of the personal touch, has been a toxic mix for the practice of medicine. It has led the public to seek alternative treatments, which in the right circumstances can help, but in the wrong can harm. The plethora of therapies that have sprung up to fill the vacuum, where science and academic medicine have failed the public, have caused confusion for the consumer, who must try, without expertise, to make a judgment on the validity of each cure. But patients, the very people for whom medical research is intended, disillusioned with the narrowness of scientific thinking, have insisted on a wider view. And by their embrace of alternative practices, they are pushing science from without. At the same time, equally widespread advances in technology, enabling researchers to track the pathways of mind and immunity, are finally overcoming the resistance of the scientific community to the notion that the brain and the immune system can and do communicate. In this melee, scientists in general are finally being roused to the necessity of integrating research findings, and in medicine specifically, of doing so with the whole person in mind. But without recent advances in the sciences of mind and body—in neuroscience and immunology, in endocrinology and psychiatry and rheumatology—it is likely that scientists and academic physicians would still be where they were a decade ago, mired in the focused, narrow thinking of seventeenth-century rationalism.

Because the assumptions of rationalism are so integral to scientific thinking, it is not surprising that it took the development of sophisticated technologies to convince scientists, through the language of science, that such mind–body connections are real. Until the last decade we simply did not have the tools capable of demonstrating the physical and molecular underpinnings of both emotions and disease. And without these, we could not begin to understand the biological basis of the healing power of the ancient Asclepions. In the absence of measurable physical evidence, the effects of such unquantifiable forces on health were dismissed as imaginary. But medical science is now at a crossroads of discoveries that finally allow us to piece together the mosaic of the biological basis and physiological effects of sleep, relaxation, and even prayer. We can finally begin to understand, and therefore believe, what the ancient priests knew intuitively about the curing effects of these activities. And by understanding these connections in modern terms, in the language of molecules and nerve pathways, electrical impulses and hormonal responses, scientists can finally accept that such effects are real.

—◀o▶—

Think for a moment of what it is like to be sick. The boundaries of your world shrink to the edge of your mattress. Light filters through half drooping eyelids, and the cool sheets rub against your hot skin. A bowl of soup sits abandoned on the nightstand. The gulf of floor between you and the bathroom door seems as daunting as a desert trek. You will yourself to get out of bed, but each muscle aches and the weakness is overwhelming. You fall back against the pillows, exhausted, and demoralized.

Illness can be teased apart into its discrete components: fever, fatigue, sleepiness, weakness, sadness, loss of interest in the environment, loss of appetite for food and sex, and an overwhelming desire to be still. Each of these feelings can be explained by the effects on the brain of various molecules released from immune cells during an infection. But we usually describe all these components with one parsimonious phrase: "feeling sick." These two words compactly convey the notion that our awareness of being ill has a sensory component, such as pain, and an emotional component, such as feeling sad.

The notion of feelings as an integral part of illness is universal—not only across continents and peoples, but across the divide of time that separates us from our ancestors. For thousands of years, humankind has been fascinated by the apparent connection between emotions and disease. The link that connected early peoples both to their mysterious, unexplainable, often threatening outside world, and to its apparent precipitation of disease, was their emotions. Even before the Greeks were ministered to in the soothing surroundings of the Asclepion, the belief that emotions could cause disease held sway in popular culture. We encounter it as far back as the Hittite king who developed paralysis and speech loss after a frightful thunderstorm, or in the fifteenth-century belief in monster births if a pregnant woman was exposed to a fright. This belief carried through to the nineteenth century and, in transmuted form, on into the twentieth century, in grandmother's warning that if mother spilled coffee on her leg, baby would be born with a birthmark on the thigh. Although such magical thinking, rooted as it often was in the academic teachings of an earlier time, never disappeared from the popular culture, the professionals who dealt with illness tried to explain that emotion–disease connection using their best available tools. As technological advances made new investigative methods available, different aspects of this relationship were explained in one fashion or another. Some explanations turned out to be dead ends and eventually fell by the wayside, to be taken up again in the realm of popular culture—something grandmother told you that you didn't want to admit you believed. But other explanations held up to the scrutiny of scientific proof and were removed from the realm of the magical, to be ensconced in the doctrines of medicine.

The central principle of medical teaching that for a thousand years linked emotions and disease was the balance of the four humors: blood, yellow bile, black bile, and phlegm. These visible secretions were physicians' only window into the workings of the body. Imbalances in them were equated not only with disease but also with emotions. Vestiges of the concepts are buried in the words we still use to describe emotional types: sanguine, melancholy, phlegmatic, choleric. *Sanguine,* from the Latin *sanguineus* for "blood," describes an optimistic,

Since ancient times, the central principle of medical teaching was the balance of the four humors: blood, yellow bile, black bile, and phlegm. Imbalances in them were equated not only with disease, but also with emotions. It was believed that these imbalances could be detected by subtle differences in the color of the urine—hence surrounding the descriptions of personalities are flasks, shaded in the original hand-colored edition of this engraving from a 1495 medical textbook by Johannes de Ketham.

confident person. In the 1495 *Manual of Medicine* by Johannes de Ketham, a sanguine person was described as fat and merry and liking Bacchus and Venus, the gods of wine and love. Not a surprising description perhaps, since these conditions—drink and love—are often associated with a rosy or blushing countenance, which is indeed caused by blood rushing to the cheeks. The opposite type in de Ketham's text, the melancholic, is a combination of *melan*, Latin for "black," and *choler*, or bitter bile. A melancholic person is gloomy and bitter. But pure bile, or choler, makes one impetuous and irascible. Today, the French word for *anger* is *colere*, and the root of the word shows up also in a "colicky" baby—one who is irritable. Phlegm, on the other hand, makes one fat and languid, slow-moving. Today *phlegmatic* has come to mean stolidly calm, unexcitable, and unemotional.

Coursing through each of these words and their modern deriva-tives is a simple assumption that tells us a lot about the differences across time in the people who chose them to denote phenomena they observed. In medieval times, the humors conveyed the idea of visible, physical bodily secretions, inextricably linked to emotional responses. But over the centuries, the physical notions fell away and only the emotional elements of the words remained.

Among the forces that led to this excision of the physical from the emotional in our words and in our scientific thought were new principles of medicine. These principles grew out of the discoveries of the sixteenth-century anatomists who had begun to dissect the human body. Thus the prevailing doctrine of the balance of the four humors began to give way to the hard, visible proof of anatomy. Dissection showed that there were normal anatomical forms and connections of organs within the human body. In disease, these forms were distorted, often riddled with holes, filled with pus or blood. Disease was then defined in terms of abnormalities of anatomy. But there were many diseases in which anatomical dissection did not reveal an abnormality. These were usually diseases of emotions and the mind. In the anatomists' terms, how could these illnesses be real?

By the end of the nineteenth century, the anatomists' notion that all disease had a physical basis, so heretical in the fifteenth century, had become so much a part of the dogma of medicine that now the heretics were those who insisted that even diseases of the emotions, with no discernible anatomical basis, should be taken seriously. A struggle arose, which continues to the present day, between the physicians of the mind and those of the body—as if mind were somehow not part of body. Those concerned with the study of the mind, disillusioned with the power of anatomy to explain all illness, came up with an alternative term for diseases that could not be explained by abnormal anatomy: the "functional neuroses"—"functional" because no structural change could be found in the brain, yet no less apparent, no more difficult to diagnose than "real" disease. This discipline was rooted in the psychoanalytic theories of doctors like Sigmund Freud, himself a neuroanatomist frustrated by the inability to identify an anatomical cause for illnesses like hysteria. Freud developed his theories of psychoanalysis, based on the available knowledge of the day, to fill the void and to try to explain, in his words, "the puzzling leap from the mental to the physical." So the rift widened between those who studied the mind and those who chose the easier, more concrete task of studying what could be seen: physical illnesses of the body.

In the late nineteenth and early twentieth centuries a new discipline arose in an effort to explain illnesses with no visible cause: psychosomatic medicine. Although now the term "psychosomatic" carries with it close to a century's worth of baggage—associations suggesting hypochondriasis and diseases that do not really exist—the term was originally meant to encompass diseases of the body (*soma*) caused by the spirit, or soul (*psyche*). Psychosomatic medicine was eventually applied in attempts to explain many physical ills in terms of the underlying psychological disturbance. The doctors who developed this school of thought, Helen Flanders Dunbar and Franz Alexander, claimed a psychological source for illnesses such as asthma, arthritis, heart disease, and gastric ulcers. But because they lacked the tools to prove such a connection, their theories fell into disrepute with the academy, which had moved on to demand stringent proof that such connections existed at a molecular and physiological level.

The teachings of psychosomatic medicine filtered into the popular culture, however, and stayed there. They joined existing beliefs about mental influences on the body and echoed back repeatedly in countless magazine articles and self-help books, even in popular song. In "Adelaide's Lament" from *Guys and Dolls,* the thickly accented Adelaide sings ruefully,

> In other words, just from waiting around for that plain little band of gold,
> A person
> Can develop a cold.
> You can spray her wherever you figure the streptococci lurk,
> You can give her a shot for whatever she's got
> but it just won't work.
> If she's tired of getting the fisheye from the hotel clerk,
> A person can develop a cold.

In modern medicine we have no difficulty in distinguishing between physical organs and the complex functions they perform, except when it comes to emotions and thought—the functions of the brain. These functions are so complex, so hard to grasp in physical terms, and until recently so hard to relate to the underlying nervous system structures, that the functional illnesses of emotions—the extremes of feelings that most of us have experienced in some measure at some time in our lives—could too easily be dismissed by those whose framework of proof was rigidly tangible.

What was missing at the end of the nineteenth century to explain the connection between emotions and disease was a knowledge of how these complex phenomena occurred at a level beyond what could be seen with the naked eye. It required discoveries of mechanisms that we cannot see, not with our eyes, nor even with the microscopes that did reveal to the scientists of that era the histological basis of disease. Seeing through a nineteenth-century microscope those nerve cell connections and immune cell infiltrations, or even seeing with a more powerful twentieth-century microscope the tiniest organelles within these cells, still didn't explain how these organs and cells could produce emotions nor how disturbances in them could cause disease.

It took very different kinds of technology—chemistry, biochemistry, cellular and molecular biology—to help us see the ways in which the nervous and immune systems work and communicate. We can now see that those cells—which to early microscopers appeared as immobile ghosts, fixed in their tissue firmaments—actually crawl about, meet, fuse, divide, and grow. Beneath our skins is a constantly changing world where our body's cells fashion and refashion our solid-seeming tissues. The mortar that they use to accomplish this is made of chemicals—amino acids, linked together in long chains to form proteins, which in turn form scaffoldings outside and inside the cells. These proteins are made in factories within the cell, factories we can now see with powerful electron microscopes, factories that appear like shelves and budding sacs. But even these solid-seeming structures are made of thousands of proteins and other molecules, linked together in perfectly fitting mosaics. And the directors that program all this activity are the genes within the cell's inner core. So now, too, we can actually see the DNA that makes up these genes, we can see its shape and where the molecules that control its function sit within its coils. With this knowledge, we can finally see how tightly the nervous and immune systems are linked through many interwoven strands of nerve pathways and communicating molecules. And once we understand that, it is not so difficult to imagine that forces that might perturb one system would have powerful effects on the functioning of the other.

We have the tools, then, to see beyond the limits of the microscopic images of the histologists of the nineteenth century, deep within the cell, to the very genes that make it function. At the same time we can move beyond the limits of the outlines of anatomy to see, with computer imaging techniques, the living human brain at work. By combining these tools we can understand how the brain receives signals from the outside environment and how these signals are processed into perceptions as well as emotional, physiological, and hormonal responses. And with advances in cellular and molecular biology, we can piece together how such nervous system and hormonal changes can affect our susceptibility to disease. By parsing these chemical intermediaries, we can begin to understand the biological underpinnings of how emotions affect diseases: viral or bacterial infectious diseases, such as AIDS and tuberculosis; inflammatory diseases, such as rheumatoid arthritis and lupus; and illnesses in which extreme fatigue goes along with aching muscles and joints, such as fibromyalgia and chronic fatigue syndrome.

Through this very research, scientists are finding that the same parts of the brain that control the stress response, for example, play an important role in susceptibility and resistance to inflammatory diseases such as arthritis. And since it is these parts of the brain that also play a role in depression, we can begin to understand why it is that many patients with inflammatory diseases may also experience depression at different times in their lives. Thus, the psychosomatic notion that inflammatory and allergic diseases originate in a disordered upbringing and repressed emotions can now be reexamined in more precise physiological terms. Rather than seeing the psyche as the source of such illnesses, we are discovering that while feelings don't directly cause or cure disease, the biological mechanisms underlying them may cause or contribute to disease. Thus, many of the nerve pathways and molecules underlying both psychological responses and inflammatory disease are the same, making predisposition to one set of illnesses likely to go along with predisposition to the other. The questions need to be rephrased, therefore, to ask which of the many components that work together to create emotions also affect that other constellation of biological events, immune responses, which come together to fight or to cause disease. Rather than asking if depressing thoughts can cause an illness of the body, we need to ask what the molecules and nerve pathways are that cause depressing thoughts. And then we need to ask whether these affect the cells and molecules that cause disease. In order to find the answers to these questions, we must know something about the brain's nerve pathways that underlie emotional responses and also about the body's normal immune responses to chemical or physical threats.

Since new research is showing that inflammation is an important part of diseases previously thought not to have an inflammatory basis—for example, Alzheimer's disease and heart disease—understanding the brain pathways that predispose either to susceptibility or resistance to inflammation will open up new ways to successfully treat an even wider range of illnesses than previously imagined. We are even beginning to sort out how emotional memories reach the parts of the brain that control the hormonal stress response, and how such emotions can ultimately affect the workings of the immune system and thus affect illnesses as disparate as arthritis and cancer. We are also beginning to piece together how signals from the immune system can affect the brain and the emotional and physical responses it controls: the molecular basis of feeling sick. In all this, the boundaries between mind and body are beginning to blur, even for the academics who had for so long worked to keep those boundaries sharp. These questions can now be answered at a level of detail that will offer new treatments for disease and new ways of interacting with the stresses in our environment that we cannot control.

CHAPTER · 2 ·

Where Do Emotions Come From?

————◄o►————

*T*he juxtaposition of the words "emotions" and "disease" is both jarring and tantalizing. Emotions are amorphous and uncontrollable. Disease is concrete and identifiable. We identify diseases in terms of what they look like, their signs and symptoms. We define emotions in terms of what we do and how we feel. A disease doesn't do anything. It simply is. An emotion makes us do something. It makes us feel. Emotions are what we live and die for. Diseases are what we die from.

Diseases are easy to locate—we can point to the place where it hurts or to some spot or bump or bruise. "Chickenpox" evokes a child with a bumpy skin rash; "pneumonia," someone with fever, coughing, and chest pain; "arthritis," a person with crippled hands. But emotions seem to bubble up from some ill-defined location within us. Happiness evokes smiling and warmth; sadness, crying and emptiness; anger, yelling and raging. Emotions can be discrete or can blend seamlessly together in endless shades. Diseases can mix but maintain their discreteness.

Emotions are always with us, but constantly shifting. They change the way we see the world and the way we see ourselves. Diseases come and go on a different time scale. And if they change the way we see the world, they do it through the emotions. Could something as vague and fleeting as an emotion actually affect something as tangible as a disease? Can depression cause arthritis? Can laughing and a positive attitude ameliorate, even help to cure, disease? We all suspect that the answers to these questions are yes, yet we can't say why and certainly not how. Indeed, entire self-cure industries have been built on this underlying assumption. But physicians and scientists until recently dismissed such ideas as nonsense, because there did not appear to be a plausible biological mechanism to explain the link.

Is there something about the biology of emotions and disease that gives them their different characteristics? Is it something in this biology that allows one to affect the other? These questions arise at the intersection of popular belief and everyone's own personal observations that emotions have something to do with disease. The disconnect that then occurs between these questions, which grow from the essence of our human experience, and the lack of concrete explanations that satisfy rigorous standards of proof has led to a mistrust between the questioners and the scientists who are expected to answer them. Part of the reason for this is that scientists and lay people speak different languages—but so do emotions and disease. Poetry and song are the language of emotions; scientific precision, logic, and deductive reasoning are the language of disease.

It is probably not a coincidence that we use the word "feeling" to describe two very different and yet very connected notions: how we feel and what we feel. The phrase "feeling sick" implies both a sensory and an emotional component. And the phrases we use to describe our emotional states also contain a physical assumption: we "feel" sad, anxious, fearful, or happy. The word "sensual" actually contains within its root the Latin word for "touch," or "sensation"— *sensus*. If we want to describe strong emotions, we use the word "palpable"—something we can touch. And we feel the sensations that we associate with emotions in distinct parts of our bodies. Our heart flip-flops in our chest, and we feel anxious; a gnawing in the pit of our stomachs signifies fear; a warm spreading feeling in our belly, and we feel happy. We don't actually feel any of these emotions in our heads. So it is no wonder that until relatively recent history, it was not at all apparent to physicians that the seat of the emotions is in the brain. Rather, logic suggested to the medieval guardians of knowledge that

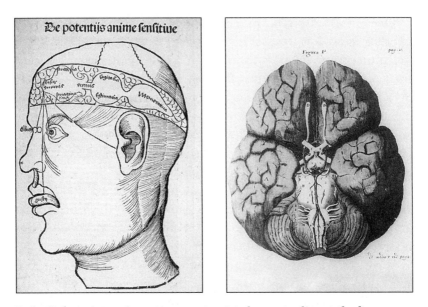

Left: *Before sixteenth-century anatomists began to dissect the human body, scientists and physicians had only a vague notion of what the brain looked like, as shown in this 1517 woodcut from Gregor Reisch. Right: Once anatomists began to explore the interior of the human body, accurate representations of the body and its organs, such as this 1679 etching by Thomas Willis, were produced.*

the seat of the emotions was in the liver or the heart. Hence the importance placed on the humors bile and blood.

◄o►

Before the sixteenth century, physicians only had a very vague idea of what the brain and other internal organs looked like, largely because the church sternly prohibited the dissection of the human body. But in the sixteenth century in the great universities of Europe, a counterculture of sorts arose and flourished, made up of anatomists and academicians willing to risk the wrath of the authorities. These brave early explorers began to chart this new territory, with the result that the inner landscape of the body was gradually laid out in perfect detail.

One of the most fertile of these centers of anatomical pursuit was in Padua, today less than an hour by train from Venice. Even in the sixteenth century, Padua was less than a day's journey by horse or boat from the seat of this fiercely democratic republic, ruled by the doges and their elected parliament from their pink and white, faintly Arabian-looking palace on the edge of the Venetian lagoon. In the early sixteenth century, the University of Padua was already more than 300 years old. Within the jurisdiction of the Republic of Venice, the university enjoyed, and even flaunted, some independence from the strict rulings of the church. But both also flourished side by side. The university and the great Basilica San Antonio dedicated to the patron saint of healing, St. Anthony of Padua, were both founded in the thirteenth century, just a decade apart.

Some historians say that this was not a coincidence. The first wave of students who flocked to Padua had followed the healer and their teacher, Anthony, from the older university at Bologna, and many more migrated there and stayed there when he died less than a decade later. More than three centuries later, in 1595, as the last stones were being laid to complete St. Anthony's cathedral, the medical school of the University of Padua, a few hundred meters away, was also still being built—constantly renovated and expanded to accommodate new students. Within the cool dark interior of the cathedral, under Romanesque domes painted like a starry sky, the lepers and the lame came to be healed. They came from across Europe to pray to Sant' Antonio, Saint Antoine, Saint Anthony, leaving votives behind in gratitude for their newfound health. Today the faithful still come on pilgrimage, carrying children to be blessed, wheelchair-bound patients, or photos of sick loved ones. The votives, or "grazie," they leave behind near St. Anthony's tomb are hearts and limbs made of tin or silver, and candles.

When you emerge from the cathedral into the glinting sunlight, you can cross the square past a short alleyway that passes over a narrow canal and dead-ends at a pair of iron gates set in a brick-and-stone wall. The tall trees and pebbled paths outlining concentric rows of tended gardens behind these gates are the botanical gardens of the University of Padua. These were also founded in the sixteenth century, by an edict from the Senate of the Republic of Venice, to house and allow study of precious and medicinal plant specimens that explorers had brought back from the New World, Asia, and Africa: palm trees, bananas, goldenrod, marigold, and aloe. Crossing back into the square, and then into the shade of a patchwork of sidewalks, covered by cool stone archways, you can hurry through the maze of streets, past the market square, to the tall gates of the main building of the medical school, the Palazzo del Bo: the Palace of the Ox.

As the gates swing closed and clang shut, you leave behind the noisy square—filled today with tourists, shops, and cafes—and enter the courtyard of the world of the sixteenth-century anatomists of the University of Padua. The two-story square building, with a red tiled roof and ochre stucco walls, that surrounds the courtyard, lined on each level by a colonnade, was said to contain apartments for the wealthy before the thirteenth century. By the time the university was founded, it had become an inn, the Albergo di Bo: Inn of the Ox. This was where the group of renegade students, disenchanted with the academic restrictions of the old university in Bologna, chose to meet and study, under the protected freedom of the Republic of Venice. Eventually the Republic renovated the structure and converted it to house the university, but kept its name, calling it Palazzo del Bo. The words on the university's emblem still reflect the freethinking spirit of its founders: Universa Universit Patavina Libertas—Here in Padua all are free to study all.

If you skirt the flagstone courtyard to the left, you reach a second set of iron gates, which shield a staircase at the corner. Passing through these gates, you climb up worn and shiny limestone stairs to the second-story colonnade. Here you can lean over the balustrade to see the courtyard below. If it is the time of year when students are receiving their degrees, or "Laurea," you can watch the same raucous celebrations that have reverberated on these stones for hundreds of years. Students draped in laurel wreaths (and little else, save underwear) run the gauntlet of their classmates, as they are literally kicked out of school. Today the professors barely tolerate such behavior. In 1595, it would have been Professor Hieronymus Fabricius di Aquapendente who swept past this balustrade, frowning at the antics below. But now only his marble bust adorns the portal to the hall of medicine he would have entered. Fabricius was the university's first professor of anatomy, whose vision it was to build a permanent anatomical amphitheater in which to teach students the dissection of structures of the human body.

Passing through the portico, you find yourself inside an oblong room with a heavy oak crossbeamed ceiling decorated with traces of color—reminiscences of the painted ceiling from when the chamber was an inn. The stucco walls are now lined with gilt-framed oil paintings of the professors who lectured to countless generations of medical students within these walls—paintings commissioned in the eighteenth century by Professor Morgagni, yet another great name in the annals of anatomy. You can almost hear the swishing velvets and silks of their robes as they moved through the throngs of students to the lecture chair—Harvey, Vesalius, Fallopio, Morgagni, Willis, and, of course, Fabricius. Medical students will recognize these names as organs of the body—each professor having left leaving his name on the structure he discovered: the Fallopian tubes of the uterus; the foramen of Morgagni, the hole in the skull through which the brainstem passes; the circle of Willis, a circle of blood vessels at the bottom of the brain; the bursa of Fabricius, an immune organ in birds. And anatomical textbooks through recent decades still contained the exquisitely detailed etchings of William Harvey, whose studies elucidated the entire circulatory system, and of Vesalius, whose classical series of drawings showed peeled-back layers of human anatomy from skin to muscles to bones.

To the right of the wall of portraits you are facing, at the narrow end of the oblong room, just to the right of the door you have entered, are two more small doors, the only other apparent exits from the room. These doors are wooden and unassuming—no archways or busts decorate them. They could be doors to broom closets. If you use a large skeleton key to unlock the door on the right and step through it, just as the medical students of the sixteenth century might have done, you face a curving stone staircase that winds around a curved wooden wall. This structure, which looks, from where you stand, much like the wall of a huge wooden bucket, is the outer wall of the anatomical dissecting theater. At the time of the anatomy lesson, the students would have poured through the door and scrambled up the staircase, squeezing through small trapdoors, to take their places along the wooden rows of the amphitheater.

Peer into the shadows, and you can make out another door that leads off to the right, to a square stone antechamber, not larger than a modern living room. Here in sixteenth-century Padua, the surreptitiously obtained human corpse was readied for the anatomy lesson,

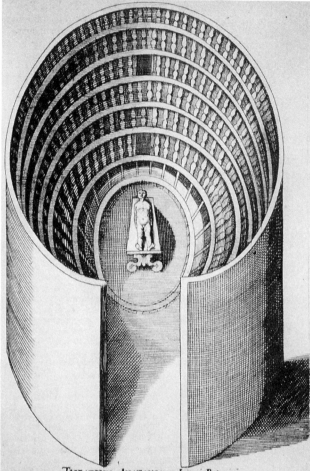

THEATRVM ANATOMICVM Lycei Patauini.

Although discouraged by the church, human anatomy lessons were held, nevertheless, at the University of Padua, one of the great European centers of learning in the sixteenth century. After being readied for the lesson, the surreptitiously obtained corpse was then placed onto a slab at the bottom of the University's amphitheater. This systematic dissection of the human body led to an era of great discovery of the detailed structures and relationships of the internal organs and how they might work. It also led to a recognition that many of the normal forms of these organs become abnormal or are destroyed once affected by disease. It is not too strong a statement to say that such dissecting theatres changed the course of modern medicine and refocused the science of medicine on a solid basis of exploration and observation.

having been held there until the moment came for it to be lifted onto the slab at the bottom of the amphitheater. Doubling over to pass through this low opening through which the corpse would have been slid, you can look up into the steep oval amphitheater. It is a view that only the professor, in his thronelike chair, and his few assistants would have had, for the small wooden floor was large enough to hold only the body on its slab and these few teachers. The students thus lined the railings of the amphitheater, straining over the five tightly packed steep concentric rows to see the professor in his rich, thick robes demonstrate the dissection of the human body.

At the start, this was a risky endeavor, fraught with danger of retribution from the authorities of the Church. Indeed, although the truth of the story is still debated, legend tells that sentinels stood watch to warn of the approach of the authorities. At the first sign of such an intrusion, the trapdoor underneath the slab was pulled open, allowing the body to slip away into the roiling river, now extinct, that then flowed directly under this hidden amphitheater. The professor was left innocently in his chair, lecturing alone to his attentive students. It has been said that the anatomical dissecting amphitheater in Padua was built as it was to facilitate such clandestine activities, thus allowing these brave explorers of the body the freedom to carry on.

The movement toward systematic dissection of the human body that these few began led to an era of great discovery of the detailed forms, structures, and relationships of the internal organs, and from there to a deeper understanding, by deduction, of how these organs might work. It also led to a recognition that there are normal and abnormal structures of anatomy and that, in disease, many of the normal forms are abnormal or destroyed. It is not too strong a statement to say that Fabricius's vision and determination to build a permanent dissecting amphitheater changed the course of modern medicine, placing the science of medicine on a solid basis of exploration and observation.

So it was noticed that if you dissected the muscles of the human arm, you found sprouting from them glistening white branches— nerves that then flowed together into sinewy trunks. You could trace these bundles of nerves from the muscles, all the way up through the fat and lymph nodes of the armpit, back through the muscles of the neck, to the bones of the vertebral column. And if you opened up those bony arches, you found that the nerve bundles were attached

to a long pale-gray column of tissue running the length of the verte-
brae, receiving similar bundles from the legs and abdomen. This was
the spinal cord. And if you traced the spinal cord up to the top of the
vertebral column to the base of the skull, you found that it widened
and then, passing through a hole in the skull, attached directly to the
base of the brain. It seemed like all the muscles and organs in the
body, including the liver and the heart, sent similar nerve trunks to
the brain, although some did so through intermediate nodes in the
belly or the chest. And so, it came to be thought that it was perhaps
the brain, not the liver or the heart, that ruled the body.

But if you also look inside the brain, to try to see how it might
work, you are left baffled. These great nerve trunks simply end in a
gray and white mass of tissue, with many folds of the surface and
with islands of lighter and darker color on the inside—shapes that
reminded the anatomists of a sea horse (the hippocampus), or a pale
globe (the globus pallidus), or a woman's breasts (the mamillary bodies). The anatomists who discovered these landmarks identified them
according to their shapes or with their own names. For it was impossible to tell just by looking at them what many of these parts of the
nervous system did. Locked within the shapes in the brain must be
the secret to its functioning, but no amount of anatomical dissection
could unlock what that secret was. To do so would require the development of new methods of detection, capable of seeing deeper into
those shapes to understand how they function.

In the seventeenth, eighteenth, and nineteenth centuries, just as
astronomers were perfecting the telescope to allow them to see farther
into the heavens, physician-scientists were perfecting the microscope
to permit them to see farther into the body—into the organs charted
by the sixteenth-century anatomists. By the late nineteenth century,
these instruments were sufficiently powerful to allow scientists to see
the ghostlike filaments and shimmery bodies, much like water drops,
contained within these tissues. These variously shaped creatures,
compressed together to make up the tissues' bulk, were, in fact, different types of cells. But microscopes alone were not sufficient to allow
scientists to see the details of how these cells were connected, or even
the details of their shapes. Squint as they might, they could not discern how or whether the cells that bundled together in nerves were
joined. Thus, unable really to see, some early microscopists postulated
that nerve cells formed continuous tubes, through which the fluid of

thought could flow. What they lacked was some way to bring the pale, colorless ghostlike shapes into view. The tools that finally gave these scientists the ability to see the cells' outlines in sharp and vivid color came from an unlikely source: the Industrial Revolution.

This era of industrial innovation brought with it a new middle class, hungry for the niceties of life, including yards of brightly colored silks and velvets, folded and flounced into fine clothes and draperies. Chemists, many in the Ruhr region in Germany, were experimenting with chemicals that produced color. They began to produce large quantities of chemicals that when mixed with water colored the silks and velvets more vividly than the natural dyes, extracted from beetles and barks, that had been available for thousands of years. Suddenly, brightly colored dyes were readily and cheaply available, not only to manufacturers of cloth but also to microscopists. And the same property that enables chemical dyes to stick permanently to silk or wool fibers allowed histologists to use these dyes to color those faint cells they saw. These dyes bound tightly to the fatty membranes of cells when slices of tissues, fixed in formaldehyde to prevent decomposition, were soaked in them. Other dyes seeped inside the cell membrane and bound to the proteins within the cell. Still others didn't bind to cells, but bound instead to the amorphous protein-glue around the cells. Just as suddenly as the clothing of the middle class took on the bright hues of the newly available colors, the details of cells lit up in pink, blue, black, and red tapestries that could be seen through the lens of a microscope.

Histologists, such as the German Rudolph Ludwig Karl Virchow and the Spaniard Ramon y Cajal, added such dyes to the thin slices of brain or nerve on their microscope slides. With painstaking attention to detail, they were able to see that brain tissue was composed of nerve cells linked one to another through a myriad of connections. These cells had different shapes from the cells of the rest of the body's tissues. They had bulbous heads, with masses of short branches in a crown, like a head of hair, and they had one long foot that could extend many inches and ended almost invariably by touching the head of another nerve cell. If one followed these cells from head to foot and then on to the next cell, one could piece together many winding pathways leading from one part of the brain to another, through white and gray matter, through folds, and in and out of the strangely shaped bodies that the anatomists had named for their resemblance to animals and other familiar objects. It turned out that these gray areas

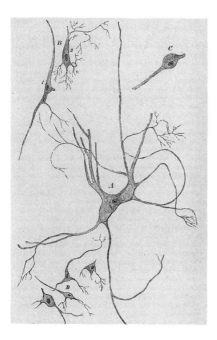

With the invention of the microscope, scientists could begin to see beyond the organs themselves and into the very tissue that composes them. By the nineteenth century, microscopes had been sufficiently refined and improved to allow Rudolf Virchow to create this 1858 drawing, which represents in perfect detail the various cells found in nerve tissue.

were really collections of large concentrations of nerve-cell bodies—switching stations in a massive tangle of wires whose reason and overall pattern of organization had yet to be discovered. But even that level of understanding didn't explain how thought and emotions arose from this mysterious organ. If the organization of these wirelike connections truly had something to do with thought, there had to be some way for these cells to transmit messages from one to another.

In the late nineteenth century, whole worlds of communication were being opened up by inventors like Thomas Edison and Alexander Graham Bell, who based their inventions on the newly discovered theories of electromagnetism. The notion that you could translate electrical impulses into coded messages, or even sound, not only allowed the development of the telephone and telegraph but also permitted a mindset in the biological sciences that was open to the idea that coded electrical signals could form the biological basis for thought and for the nervous system's command codes to the body. In the 1920s, the British neurophysiologist Sir Charles Scott Sherrington proved that nerves do in fact signal each other with moving bursts of electrical charge. Sherrington later won the Nobel prize for this discovery.

If you plunge a fine needle into one of those nerve bundles, sprouting from, say, the muscles of the arm, and attach the needle to an oscilloscope and a microphone, you can record electrical impulses passing along the nerve in machine-gun-like bursts. This method of recording nerve activity is used in clinical medicine to determine whether and what kind of nerve injury is present in patients with weakness or numbness. If you close your eyes and listen to such a recording, you can hear mysterious rhythms of staccato bursts. The number and strength of these bursts change as the patient flexes and relaxes the muscle to which the nerve is attached, sometimes reaching a whining fever pitch as the trains of electrical impulses come so thick and fast that individual blips can no longer be detected out of the din.

But unlike the dry wires in telephone trunk lines, those made of the spidery nerve cells float in a bath of liquid whose salt concentration is very much like the seawater from which our primordial ancestors sprang. It is the salts in this primordial soup that allow the electrical impulses to arise. This occurs because the membrane of a nerve cell is arranged very much like a transistor—in sandwichlike fashion, allowing salts of opposite electrical charge to build up on either side of the membrane. As the build up reaches the point of bursting, it can suddenly discharge, emptying the accumulated electrically charged salt molecules from one side of the membrane to the other. Many things can trigger this sudden avalanche, and once set off, the charge moves rapidly along the membrane like a falling row of dominoes until it reaches the end of the nerve's foot process. At this terminal shore, the charge could end and dissipate if it had no way of traversing the gulf of fluid between it and the next nerve cell. But here the nerve process balloons out into a widened lake, filled with tiny sacs of chemicals—neurotransmitters. When the charge reaches these sacs, they burst and empty into the surrounding fluid, and eventually, though more slowly than the charge itself, they reach the opposite shore. Here they trigger a whole new avalanche of discharges in the next nerve cell. And so the message is transmitted from cell to cell, on up to its final destination in the brain, if the nerve is a sensory one, or down from the brain, if it controls movement.

The triggers that can set off these gun bursts of activity in a sensory nerve cell are the thousands of stimuli to which we are exposed during every moment of our lives. They are the light touch of a cloth or a breath of wind on our skin. They are the soft warmth of the sun, or the flaming heat of burning coal. They are the shape and color of a bird whose image falls on our retinas; the pulsating sounds of drums or the soothing sounds of a waterfall; the taste and smell of fresh-baked bread or the telltale scent of perfume. We receive all these sensations through different specialized organs—magnificently engineered instruments designed to collect signals of all these different types and transduce them into a common code of electrical impulses. It is through such sensory organs—the eyes, the ears, the nose, the taste buds, and the skin—that our brains keep constant vigil on the ever-changing world around us. And it is through the tangle of nerve pathways along which the impulses run that the brain and nervous system can receive incoming signals and then direct the body to respond to each stimulus.

If it was the anatomists who led the way in the sixteenth century, it was the physiologists who, in the late nineteenth and mid-twentieth century, began to try to explain disease in terms of the body's responses to external stimuli. Physiologists, such as the Frenchman Claude Bernard, in the 1860s, and the American Walter B. Cannon, in the 1930s, found that you could dissect out a whole muscle from a frog's leg, suspend it carefully from two threads in a bath of salt water through which oxygen is bubbled, and it would still contract. You could then record its contractions with a needle, which scratched out a wavy trail on a smoke- and wax-coated cylinder. The white line that was etched, like a turn-of-the-century phonograph recording, registered the muscle's contractions in response to electrical stimulation or to whatever unknown chemicals were added to the bath.

So these scientists made extracts, like a kind of broth, from all sorts of organs, added them to the bubbling salt water, and recorded the contractions of the suspended tissues: frogs' leg muscles, pieces of rodents' blood vessels or intestines. Once an extract was found to have an effect, its chemical constituents were isolated and named for the kinds of contractions they produced: long, slow, and languid or rapid, short, and sharp. Many of these compounds, to which they gave names such as serotonin or adrenaline, turned out to be small molecules derived from the building blocks of proteins, the amino acids. And many of these turned out to be the chemicals contained in those sacs at the ends of nerves that, by emptying, carry electrical signals across the gulf between nerve cells. They called these compounds neurotransmitters.

Emotions, and diseases of the emotions, could finally begin to be thought of in terms of something concrete: perturbations of these

neurotransmitters. Claude Bernard and later Cannon and his colleagues were now able to formulate a theory of balance and harmony
in modern terms. Rather than as a balance of humors however, the
ancient theory of what constitutes emotions was reincarnated as a
balance of neurotransmitters and hormones.

—◄o►—

Arching over University Street, on the slopes of Mount Royal in the
Canadian city of Montreal, is a covered stone bridge with slitlike
windows not unlike Venice's famed Ponte dei Sospiri, the Bridge of
Sighs. Crossing this bridge, which connects the old turreted Royal
Victoria Hospital to the Montreal Neurological Institute, you leave
the fluorescent-lit, noisy outpatient clinic and enter a quiet world.
Built in the 1930s and 1940s—an era that reflected an unquestioning
awe of the power of medicine and science in the solid and imposing
architecture of the day—the institute was developed for the study of
surgical treatments of the brain. Under the slightly yellow light of the
torchiere lamps on their black-wrought iron bases, you can see spidery drawings of nerve cells decorating the ceiling: three of the six
cell layers of the cerebellar cortex, after a drawing by the Italian neurologist Camillo Golgi. Floating in the center of the ceiling is the
image of a ram's head. This is Aries, the astrologer's symbol of the
brain. Surrounding it are four odd-shaped symbols: an eaglelike bird,
a curved staff, something that looks like a poppy pod on a stem, and
a feather. These hieroglyphs, taken from an Egyptian treatise of 3000
B.C., are said to be the first representation of the word "brain." An
ellipselike ribbon, the shape of a human brain, surrounds this con-
glomeration of symbols and contains five words spelled out in the
original Greek. Translated from a tract written by the Roman physician Galen commenting on the teachings of his Greek predecessor
Hippocrates, the words contradict Hippocrates' thesis that wounds of
the brain are always fatal: "But I have seen a severely wounded brain
healed". Just below the ceiling, curling around a gilt art deco cornice,
are the names of neuroanatomists, physiologists, and psychologists
whose discoveries led the way to modern neuroscience: Thomas
Willis, whose anatomical dissections at Padua led to a perfect definition of the blood vessels of the human brain; Sir Charles Sherrington;

the histologists Camillo Golgi and Ramon y Cajal, whose studies defined nerve cells and their microorganelles; the physiologist Ivan Pavlov, who first discovered the principles of conditioning; the neurologists Jean Charcot and John Hughlings Jackson; the physiologist Claude Bernard; the endocrinologist Harvey Cushing; and neuropathologists Franz Nissl and Alois Alzheimer. In the place of honor above a carrara marble statue of "Nature Unveiling Herself Before Science," the name Wilder Penfield was added after his death in 1976. Penfield was the neurosurgeon who founded the institute to which this hall is the entrance. One of the pioneers of modern neurosurgery, he was one who had discovered and then taught others how to help the injured brain to heal.

Deep inside the hospital, beyond the quiet hallways and patient wards, behind a nursing-station desk, lies a narrow, unassuming-looking door. But if you enter this door, as did Penfield's students in the 1930s and 1940s, you find yourself at the foot of a dark and narrow flight of stairs. At the top of these stairs is a tiny half-shell amphitheater. Built in three steep concentric rows, this balcony is very much like the dissecting theater in Padua, except that at the bottom, about 15 feet beneath its sloping glass wall and floor, is a ceramic-tiled windowless operating room. In the very middle of the room is an operating table brightly lit by hot white lights. From the 1930s on, student doctors would line the dimly lit concentric rows of the amphitheater and strain to see the surgical procedure that was taking place below. In the middle of this balcony's bottom row there is wooden trap door, just as in Padua's dissecting theater, large enough only for a man to climb through. But under this trapdoor is an enormous mahogany box camera, protected from the operating room by a wall of glass. The camera's 12-inch lens can still be aimed at a mirror just above it on the ceiling. At the push of a button, the mirror slowly moves along its track to the middle of the operating theater. In Penfield's day, a cameraman would climb inside the box and aim the camera at the mirror, angled above the operating table to reflect the patient's exposed brain. In the 1930s, the camera and the mirror were state-of-the-art for taking close-up pictures of the living human brain.

During each operation to remove scarred pieces of brain from patients with epilepsy, Penfield and his associate, Dr. Willie Cone, added to the map they were creating. But instead of a map of the brain that charted visible landmarks, as their medical forebears had done 400 years before, these neurosurgeons were creating a map that outlined the functional areas of the brain—areas that controlled movement of the fingers and toes, arms and legs; areas that controlled speech, and areas that received sensations of touch and color and sound. Penfield and Cone made these maps before cutting out the scarred seizure source in these epileptic patients in order to preserve as much normally functioning brain tissue as possible. Inch by inch, they would painstakingly insert needles into the brain, gradually approaching the seizure source. The needles were attached to electrodes, and at each point, they would allow a little current to flow—just enough to trigger the activity of a few brain cells. They and their colleague, psychologist Dr. Brenda Milner, would then observe what the patients did or ask them what they felt. The patients had to be awake for the procedure, so the surgeons could tell what parts of the brain were active and what parts were diseased. These patients felt no pain, however, because the brain, although itself the well of feeling, has no receptors for pain.

In this way, the doctors mapped out an image of our sense of touch. It is an image, literally, of a little person, lying on top of the brain; an image of ourselves, but one distorted out of proportion to what we see, something more resembling our image in a carnival funhouse mirror. They called this image of ourself within ourself a "homunculus," from the Latin for "little man." This homunculus has a very large tongue, equally large thumbs and index fingers, a small forehead, and spindly arms and legs. In fact, although not at all what we look like, this is exactly how we feel the world around us.

While the sensory pathways that terminate in the parts of the brain that form this little man are not emotions, elements of both the senses and the bodily responses they evoke are at the root of how we experience our emotions. Just as we use the same word "feeling" for either touch or emotion, we also use the word "visceral"—something we can feel in our gut. Strong emotions are gut-wrenching. This visceral component reflects the physiological responses that are put into play by emotional experiences. These are automatic responses—a quickly beating heart, sweating, or goose bumps, feelings we all associate with strong emotions. The part of the nervous system that controls these responses is called the autonomic nervous system. It was these responses that the early physiologists found possible to tease apart and measure.

But even these feelings are not really emotions. They are the physical effects of emotions. They are produced not by the higher, thinking parts of the brain but by a very primitive part of the brain called the subcortical brain at the junction of the spinal cord, brainstem, and those higher parts. The further up this anatomical chain we progress, the more abstract become the functions of the nervous system. But so grounded are we in the physical that while it is easy to conceive of and describe concrete feelings like a rapidly beating heart, sweating, or flushing, it is harder to describe or understand emotions without these physical descriptors. Try to define the sensation of fear without them. It is almost impossible—like trying to describe the color red without comparing it to blood or to a rose.

Can we experience emotions without their physical manifestations? If the physical trace that emotions leave on our bodies is not there, do emotions still exist? These questions are like the ancient question in logic: If a tree falls in the forest and we are not there to hear it, does it make a sound? And the answers to those questions are very similar. With technology we can now answer the second question in a way the ancient logicians could not have imagined; by placing a tape recorder in the forest and recording the sound, we can of course prove that the falling tree makes a noise. But if we define sound only in terms of the sensory organ, the ear, that must be there to interpret the crashing sound waves in the air, whether it is from a tape recorder after the fact or by a listener present at the time, the tree makes no sound unless there is an ear there to interpret it.

So, too, with emotions. Part of the conscious experience of emotions comes from recognition of their physical effects. One could then argue that you can't have an emotion unless you consciously experience it. But this is clearly not the case, since we can misinterpret some of our perceptions of emotions, those visceral feelings—mistaking anger for guilt, love for hate. When confronted by a controlling and demanding spouse, you might withdraw, lose your appetite, and feel a gnawing in the pit of your stomach—and think that this is guilt and sadness when in fact it is anger. Or you can believe that you have no emotional response, feel wrapped in cotton-wool insulating yourself from the world, when in fact you are deeply grieving. Sorting out these miscues from our visceral responses, diving through the layers of consciousness, is, after all, the basis of psychoanalysis.

What, then, are the components of emotion besides the physical ones? To every emotion there is a sensory end—something received by the brain through the senses that triggers the emotional response. And there is a motor end—those physical, physiological responses that we feel in our viscera, such as palpitations, sweating, increased blood pressure, hair standing on end, blood rushing to the skin. And then there is the black box in between—the places in the brain where something happens that adds emotional charge to those inputs we sense and those physical responses we recognize as emotions.

Think for a moment of the inputs and outputs alone. Every minute of the day and night we feel thousands of sensations that might trigger a positive emotion such as happiness, or a negative emotion such as sadness, or no emotion at all: a trace of perfume, a light touch, a fleeting shadow, a strain of music. And there are thousands of physiological responses, such as palpitations or sweating, that can equally accompany positive emotions such as love, or negative emotions such as fear, or can happen without any emotional tinge at all. What makes these sensory inputs and physiological outputs emotions is the charge that gets added to them somehow, somewhere in our brains. Emotions in their fullest sense comprise all of these components. Each can lead into the black box and produce an emotional experience, or something in the black box can lead out to an emotional response that seems to come from nowhere. One of the extra ingredients that can shade these physical responses and sensory inputs with feeling is memory.

Walking along, neither happy nor sad, suddenly, as if falling into a sinkhole, your mood crashes. What did it? What triggered the switch? Mood is not homogeneous like cream soup. It is more like Swiss cheese, filled with holes. The triggers are highly specific, tripped by sudden trails of memory: a faint fragrance, a few bars of a tune, a vague silhouette that tapped into a sad memory buried deep, but not completely erased. These sensory inputs from the moment float through layers of time in the parts of the brain that control memory, and they pull out with them not only reminders of sense but also trails of the emotions that were first connected to the memory. These memories become connected to emotions, which are processed in other parts of the brain: the amygdala for fear, the nucleus accumbens for pleasure—those same parts that the anatomists had named for their shapes. And these emotional brain centers are linked by nerve pathways to the sensory parts of the brain and to the frontal lobe and hippocampus—the coordinating centers of thought and memory.

The same sensory input can trigger a negative emotion or a positive one, depending on the memories associated with it. A scent of perfume can trigger a warm childhood feeling of mother returning from a night out to kiss you goodnight and tuck you in—or the aching sadness of a love affair ended. A fleeting shadow can trigger fear if the amorphous shape reminds you of childhood monsters lurking under bedposts—or longing if it reminds you of an absent lover sliding into bed.

Probably the most famous description of a sensation triggering emotional memories is Marcel Proust's description of the taste of a sweet cake in the first volume of his epic work *Remembrance of Things Past*. The mere taste of a "madeleine" soaked in tea brings on the faintest hint of a memory. After straining to resurrect the feeling, he discovers its origins to be a warm memory from childhood, of being cared for by a special aunt. Proust goes on for pages, but he begins with a perfect description of the phenomenon.

> No sooner had the warm liquid mixed with the crumbs touched my palate than a shiver ran through me and I stopped, intent upon the extraordinary thing that was happening to me. An exquisite pleasure had invaded my senses, something isolated, detached, with no suggestion of its origin. . . . this new sensation having had the effect, which love has, of filling me with a precious essence.

The emotion that Proust describes starts with a taste of cake, a taste composed of not one but a symphony of parts. The taste buds, and their close cousins the olfactory organs in the nose, are really chemical sensors that can distinguish and identify more minute concentrations of chemicals than any of the most sophisticated instruments designed by man. These organs can detect just a few molecules in a volume the size of a sugar cube.

So, as Proust's madeleine dissolves in the tea, countless molecules are gradually released into the hot liquid that bathes the tongue. Some mingle together in the fluid that coats the crevices and grooves where the taste buds lie. Others waft up the nasal passages to the olfactory organs in the nose. These sensory organs for taste and smell are constructed much like a flower. On their outer face are clusters of hundreds of cells whose surfaces are dotted with thousands of receptors—proteins whose shapes fit the chemicals they detect like a lock fits a key. When a molecule of sugar, say, or tea, finds its way into a crevice, it slides along the florescences, passing over receptors whose fit is not perfect until it slips into one whose shape is just right. Like a shot, this triggers an avalanche of chemical reactions in the cell, which in turn trigger electrical impulses in the nerve fibril that leads from each cluster. Each different kind of molecule triggers a different kind of cell, generally grouped together for categories of taste in separate areas of the tongue: sweet in the front, salty in back, and sour on the sides. So to appreciate the full taste of the tea, or coffee, or wine, you must gently roll the liquid over all these surfaces, and in so doing you can gradually distinguish new nuances of taste as cells and nerve fibers from new taste buds are slowly brought into play.

For each of the thousands of molecules released into the tea, thousands of different nerve fibers are activated. As they lead away from the tongue and nose, the fibers join together into bundles that make up the first and the last of the twelve cranial nerves: the first for smell, the last for taste. From there these nerves wind back to the stem of the brain and end in small switching stations, where new nerve fibers connect with them. These lead on to higher brain regions, the temporal lobe parts of the cortex, where electrical signals are reassembled into the tastes and smells that we recognize.

The pattern and intensity of electrical activity of these nerve fibers are an imprint of the pattern and concentration of molecules in the tea. As waves of molecules are released into the tea, waves of nerve cells become activated. So a warm feeling of mingled tastes flow over us as we roll the tea over our tongues. Not all the signals go directly to the cortical taste and smell centers. Some take short cuts to centers of memory and from there to emotional centers in the brain. And so, the waves of taste and smell are transformed into waves of emotion and memory: Proust's sensation filling him with a precious essence. But in each transformation—from chemical, to physical, to electrical impulses, to the brain's reconstruction of taste and smell, to emotions and memories—the signals lose some sharpness. What started out as clearly defined becomes a vague feeling, a warmth, a joy, whose origins seem lost in the mists of our minds.

The areas that control emotions are specialized—some for positive emotions and others for negative. This ensemble of centers within the brain, situated deep inside its core, is connected by a series of nerve pathways. In the 1930s the American neuroanatomist James Papez postulated that these centers together controlled the generation and expression of emotions. Later in the 1940s and 1950s, the American physiologist Paul MacLean termed this grouping the limbic system.

There are parts of the brain, for example, that control those positive emotional and physiological components of the thing we call love: pleasure, happiness, warmth, and comfort, as well as the set of sexual responses we associate with romantic love. This part of the brain is called the nucleus accumbens. There is also a part of the brain that controls fear. Contained on either side of the brain in two identical small oval bodies the size and shape of almonds, these structures named by the anatomists the amygdala—Latin for "almond." One clue that suggested to physicians of the nineteenth and twentieth centuries that these bodies constituted the centers of fear was a clinical pattern of behavior in patients in whom a stroke had damaged these parts of the brain. These people became docile, imperturbable, and totally without fear.

The fear center of the amygdala sends nerve fibers to those lower parts of the brain, the hypothalamus and brainstem, that control breathing, sweating, heart rate, blood vessel, and muscle tone. It also receives inputs from the parts of the brain that compose the signals we receive through our sensory organs: the visual and auditory cortexes, which register sight and hearing; the olfactory cortex which receives odor signals; and the parts of the brain that register taste and touch. The pleasure center of the nucleus accumbens it seems may also have such connections. These emotional centers, then, are wired above to incoming signals from the environment and below to motor centers, making us capable of responding to such signals at a moment's notice. It is the physiological responses programmed by the hypothalamus that prepare our body to run—increasing the blood flow to our skin and muscles, and optimizing our breathing and metabolism for the task at hand.

Through such nerve pathways, the emotion we call fear and the emotion we call love get translated into the physical sensations we associate with these feelings. This wiring allows us to respond with split-second timing almost automatically, in a fraction of a second, to the stimulus. So emotions, in a very real sense, make us move—move in the case of fear in response to stimuli that through our senses are perceived as threatening, and move in the case of love to achieve our goal and propagate the species.

An incoming sensation—that fleeting shadow—can be shunted through the amygdala and produce fear, or it can be shunted through the limbic system and produce sexual arousal, depending on many factors that we don't yet fully understand but that certainly include traces of memory of which we may not be consciously aware. If we begin to think of all these parts of emotions as coming from distinct anatomical places in the brain and realize that they are all connected by complex wiring—nerve circuits that allow inputs to lead to outputs that can be channeled through switching stations that add different sorts of charge—we can begin to think of the mysterious nature of emotions in more scientific terms. And we can begin, with the powerful tools of modern science, to map out the structures that underpin the generation of these remarkable and powerful motivators of our lives. It is the mechanism of this shunting—and the circuitry between the places in the brain that receive sensations, turn out feelings, and add charge to it all—that the newest technological advances are best poised to dissect and sort out.

Today's equivalent of the anatomists' dissecting instruments are high-power magnets, laser beams of light, or instruments that detect high-energy radioisotopes, which decay to inertness in seconds. In the presence of such detecting instruments, molecules turn into small beacons, emitting signals from inside the brain that can be detected outside the skull. The technologies to detect these molecular beacons are the tools modern-day brain explorers use to slice the brain without opening the skull or even touching a hair on the patients' heads.

Thus, computerized x-ray instruments, known popularly as CAT scanners (for "computerized axial tomography"), show cross-sectional details of anatomical slices of tissue as thin as a few thousandths of an inch thick throughout the brain and body. Positron emission tomography, or PET scanning, detects high-energy radioactive chemicals injected into the blood and used by active parts of brain. Functional magnetic resonance imaging, or fMRI, instruments detect the minute magnetic fields generated as atoms in brain tissue spin on their axes.

These instruments show changes in such molecules as areas of the brain become activated. Together with computer programs that reconstruct images from a million tiny pixels, these instruments can detect subtle changes in blood flow in minute parts of the brain—in those tiny gray animal-shaped areas that the anatomists named 400 years ago. And so we can now actually see in real time the parts of the living human brain that become active or inactive when we are happy or sad, anxious or calm, aroused or fearful. An even newer technology called optical imaging uses beams of light and changes in light scatter in cells on which the light is shone to detect minute changes in swelling of a firing nerve cell. With powerful instruments and computer programs to edit out background noise, this new technique can detect the brain cells that swell with water as they become active, and then shrink again as they quiet down. So with this technique scientists will be able to see in humans—and can now see in animals—not only larger areas of brain activity but also single cells handing off activity to other cells in circles and pinwheels, like dancers in a Virginia reel.

What these studies are telling us is that many different parts of the brain are tightly linked and work together to produce the constellation of perceptions, feelings, and actions that constitute our emotional responses. Studies using such instruments show, for example, that different parts of the brain become active when we look at sad faces or at happy faces or think of happy or sad events or relationships. And it turns out that the brain is not just one amorphous mass, like a homogeneous ball of modeling clay. It is more like many different balls of different-colored clay, gently squeezed together in one larger ball, tight enough that the pieces hold together but loose enough that they can easily come apart. The sixteenth-century anatomists knew this—they could tell this when they dissected those odd-shaped parts of the brain that they had named for familiar objects and animals. Those different nuclei lifted out easily from surrounding brain tissue, much as nutmeat from the shell. But until the advent of modern imaging technology, knowing this did not tell what these discreet, tiny brains-inside-the-brain did. Today, with imaging technologies, we can see each of these centers light up in different sequences, like programmed flashing Christmas lights. Depending on the type of stimulus—something seen or heard or thought—and the emotion evoked, from fear to joy, the signal bounces back and forth from each separate structure in the network in dizzying patterns. The emotions that we ultimately feel grow from the almost infinitely possible combinations of activation of these different brain regions in space and time.

Together, all these studies tell us that the anatomical and cellular organization of the brain's structures, which receive sensory inputs from the environment and transduce them to physical responses, do shape the characteristics of those ephemeral sensations we call emotions. It required a whole new science—the science of immunology—to discover how the biological underpinnings of emotions differ from those of the immune system, which responds not to intangible threats or temptations perceived and anticipated by the sensory organs but to the tangible present danger of chemical or living invaders. And it required an equally detailed understanding of both systems before scientists could begin to piece together the link between emotions and disease.

CHAPTER 3

The Dirty Soup Beyond Our Skin

How the Immune System Defends Against the Outside World

————◀O▶————

*W*hen the Dutchman Anton van Leeuwenhoek peered down his microscope in the 1670s and glimpsed the first "wretched beasties" teeming in a drop of water, he made an observation that forever changed our assumptions about the world around us. We do not live in a clean, clear world but in something more akin to a very dirty soup, filled with microscopic creatures of all shapes and sizes. Leeuwenhoek, a janitor by trade, was able to make these observations because, with dogged determination and precision, he had perfected the art of grinding glass lenses. With his microscopes, more powerful than any others in Europe, he could see the tiny "animals" that we have since learned populate this invisible world. There are round bacteria linked in pearl-like strands—"cocci," such as the streptococci that invade the mouth to produce strep throat; long, lozenge-shaped bacteria—the "bacilli"; and bigger organisms—the yeasts and molds and fungi. There are smaller ones, too, hardly organisms at all, just a single strand of DNA or RNA curled up inside a protein shell—the viruses. And then there are much larger

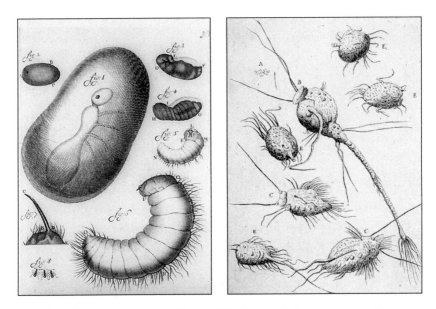

Unveiling the microscopic workings of the body's defenses began with Anton von Leeuwenhoek's perfection of the microscope in the 1670s. Ironically, Leeuwenhoek's first subjects were the very things our defenses protect us from daily—the "wretched beasties" he found teeming in a drop of water were simple bacteria.

creatures, still too small to see although we inhale them on rafts of dust: crablike with six legs and tentacles, fearsome creatures if we were less than a millimeter tall—the dust mites. Finally, as varied as these lifeforms are the pollens, which come in as many different shapes as snowflakes—some spherical, some polygonal, some spiked like the ball of a medieval mace.

The charting of this new microcosm was begun at the same time explorers were discovering new lands and astronomers were mapping the sky. It took hundreds of years, from the moment of their discovery until the late nineteenth century, for scientists to finally recognize that the denizens of this universe could cause disease. But from that time on, scientists have focused their efforts for treating disease on this world beyond our skin—learning ways to keep these invaders out and developing new chemical means to kill them. The first scientists who penetrated this universe had a hard time convincing others where these organisms came from. Debates raged over whether or not life could arise spontaneously, and if not, where it came from.

Indeed, the debate began long before microscopes were invented to see these creatures. In the first century B.C., a Roman named Varro speculated that tiny animals, so small as to be invisible, caused disease after being carried through the air and entering the nose and mouth. In 1546, the Italian Girolamo Fracastoro proposed in his treatise "On Contagion" that tiny living organisms, which he called disease seeds, could be carried by the wind or transmitted through touch and could cause disease. In the mid-eighteenth century, a group of scientists and physicians, including the British scientists John Pringle and Joseph Priestley, concluded that those illnesses characterized by chronic fevers were caused by dirt. Priestley focused on impure air and Pringle, on "septic ferment" in the blood. Even in the mid-nineteenth century, when Louis Pasteur in France and James Lister in Scotland first proposed careful handwashing and aseptic techniques for doctors delivering babies or performing surgery, their ideas were dismissed as an unnecessary waste of time. Yet women continued to die in child-birth from puerperal fever, infected by the doctors who delivered them, until Pasteur's and Lister's theories finally became standard practice in medicine.

These new treatments and preventions were so effective, and so changed the face of medicine and public health, that for decades—on into the 1960s—the focus of scientists' attention continued to be on the thing that caused the infection or inflammation, the germs, rather than the body's responses. As science evolved, the focus remained the same but delved deeper into the bits and pieces of germs that caused disease: the waxy parts of bacterial walls; the proteins and sugars that made up these waxy coats; and finally the building blocks of these larger molecules, the amino acids and the special sequences in which they are strung together into proteins that can trigger disease. But by focusing on what it is about these foreign invaders that causes disease, some scientists lost sight of the role of the body in defending itself against them.

For disease is not caused just by these microscopic invaders around us, or their pieces. It is also caused by our body's reaction to them. Think of what we experience when a raft of bacteria enter the skin on a splinter: there is redness and swelling, heat and pain, then oozing pus followed by cooling, and finally a scar. A different group of cells and tissues react to the invading organisms to produce each of these phases of inflammation that we can see and feel. Heat and redness around the splinter come from blood vessels that widen to feed the area. Swelling comes from fluid that leaks out of these vessels into the skin. Pus is the collection of white blood cells and debris that accumulate as immune cells enter the area, kill the bacteria, and die themselves. Finally, a scar forms when other cells, the fibroblasts, come in to plug the hole with collagen, a protein that acts as the glue to hold cells together in tissues. If the body didn't react in this way to a splinter, there would be no heat, pain, or redness; cells wouldn't die and need to be replaced by scar tissue. In that sense, there would be no disease in response to the splinter. But if the body didn't react this way, if for some reason the immune cells didn't recognize or respond to the invader, the bacteria that entered the skin would divide and grow unheeded. They could invade the bloodstream and cause infection in the blood and throughout the body. And the person would almost certainly die of overwhelming infection. This is what happens in immune deficiency syndromes such as AIDS. In that sense, disease, or lack of it, depends on the constant vigilance of the immune system, tuned up just right to destroy invaders before they overwhelm, but not so sensitive that it continues to act on its own once the invader is gone.

At first it was not apparent which organs played a role in defending against disease. In the days of the anatomists, certain organs were noticed throughout the body: squishy gray masses of tissue, such as the thymus in the chest and the tonsils in the throat, and firmer, rounder lymph nodes scattered inside fatty tissue in the neck and groin, as well as inside the abdomen and chest. And these organs seemed to be drained by a system of vessels, much like the veins and arteries that form the connecting waterways of the circulatory system. Except the lymphatic system, as it was called, was filled not with blood but with a clear, mysterious fluid. The function of this fluid, the "lymph," remained a mystery well into the twentieth century. And then, too, besides these white, seemingly bloodless organs, there were also red, blood-engorged organs and collections of cells: the shiny liverlike spleen on the left side of the abdomen; the bloody core of the long bones, the marrow; and odd collections of cells in what were thought to be vestigial organs such as the appendix. They were only loosely connected to each other, or not connected at all. And apart from a great duct—the largest of the lymphatics, which empties into the largest veins leading to the heart—they didn't even seem to be connected to the other organs of the body. The purpose of this loose collection of seemingly unnecessary organs strewn about the body remained a mystery for hundreds of years.

That these organs formed a single system united by a common function—what we now know to be the immune system—was not understood until relatively recently. Part of the reason for this ignorance lay in the fact that, unlike other organ systems such as the circulatory or digestive systems, if one examines these organs physically, it is not at all apparent what they do. They don't pulsate, they don't contract, they don't digest food.

Even with the advent of the microscope in the late 1600s, and its perfection by Virchow and his followers in the nineteenth century, the workings of the immune system remained a mystery. Armed, until recently, only with a microscope and some dyes of different colors, scientists looking at a smear of blood on a glass slide saw only cells of many shapes and sizes, colored differently by the dyes. There are red blood cells, colored naturally by hemoglobin, small dimpled Frisbees without dark centers, without nuclei. And there are all the rest that are not red and do have nuclei—white ghosts when there is no dye added to the glass slide to stain them. As a group, these are called white blood cells.

There are many kinds of white blood cells. Lymphocytes are the small round cells whose nuclei centers, also round and dark, take up almost the whole interior of the cell, leaving only a thin rim of pale-blue, dye-stained cytoplasm around the edge. There are bigger cells, whose cytoplasm around the nucleus is also pale blue, but whose shape is more irregular and whose nuclei are kidney-shaped. These are the monocytes. There are round cells, larger than the lymphocytes, which stain pink and are speckled with granules in their cytoplasm. They are distinguished by their multilobed nuclei, and so, because of their granules and many different—"poly"—shaped nuclei, these cells are called polymorphonuclear granulocytes. T hey are also called neutrophils because of the way they take up neutral dyes: "neuter" from the Latin for "neither" and "philos" from the Greek word for "affinity," combine to mean "having neither an affinity for basic or acidic" stains. "Cytes," from the Greek word for "cells," *kytos*, and "leuko" for white, are combined to form the word describing all these white blood cells:

leukocytes. The confusing plethora of names for these cells becomes even more confusing when one trains one's microscope beyond blood vessels into tissues.

When the first histologists looked through their microscopes, they saw many different cells in inflamed tissues—cells that in some ways did, and in other ways didn't, resemble those they saw inside blood vessels. These were of different colors and different shapes from those in the blood, so these scientists had no way of knowing that when white blood cells escape into tissues, they can change their color, shape, and size. Thus, when monocytes squeeze out into tissues, they develop long footlike extensions that help them move—these "pseudopodia," extending out in all directions, give the cells, still with their kidney-shaped nuclei, a ragged, hairy appearance. They're bigger, too. So the histologists named them macrophages—from "macro," meaning "large," and "phage," meaning "one that eats or swallows." But they had no idea that these cells were the same monocytes they had seen in blood, which had merely morphed to do their job of devouring debris left at infectious sites. The lymphocytes, too, change as they evolve to perform their task. Those small, bluish, round lymphocytes in blood mature to become antibody-producing factories, plasma cells that when stained appear in tissues as perfect pink ovals. Unlike with cells in other areas of the body, histologists could not define a lymph cell's lineage—where it came from, and what it did—just by describing its shape and color. Yet, for hundreds of years, until very recently, this is all that we could do in studying the body: look at tissues and describe what we saw.

Even more problematic are the B and T lymphocytes: it is impossible to distinguish them simply by their color and shape. Under a standard microscope, using standard dyes, they look exactly the same. The first immunologists could distinguish these cells only by their location in different immune organs—"B" for bone marrow in mammals (and for an organ called the bursa in birds) and "T" for thymus. It took decades more for scientists to discover antibodies specific to proteins on the surface of each cell type. And it took engineers and chemists as many years to develop the sophisticated instruments and microscopes, as well as the fluorescent dyes that attach to the antibodies, to detect where each bound itself. It wasn't until all this technology was perfected that these different kinds of lymphocytes could be identified and distinguished after they had left their birthplaces to circulate in blood. And once this technology was perfected, other types of experiments had to be performed to determine the different kinds of job each cell type carries out in the complex battle of the immune response.

◄O►

The great frontiers explored in the sixteenth century were those of physical space—the oceans, the continents, the mysterious and forbidding shorelines of the New World. These discoveries cast their spell on all aspects of man's creative endeavors. They can even be found in the era's literary imagery of the human body. The British metaphysical poet John Donne repeatedly used such geographic analogy. In his musings on sickness and death at the end of his life, "Hymne to God my God, in my sickness," he used the geographic metaphor to convey the continuity of life, death, and afterlife:

> Whilst my Physitians by their love are growne
> Cosmographers, and I their Mapp, who lie
> Flat on this bed, that by them may be showne
> That this is my South-west discoverie. . . .

> As West and East
> In all flatt Maps (and I am one) are one,
> So death doth touch the Resurrection.

The anatomists, too, explored the physical dimension, but instead of new continents, they delved into the world within the body.

Yet none of these explorers, geographic or anatomical, ventured into the fourth dimension: time. The rhythms of time were so much a part of their world that they took them for granted. The rising and setting of the sun, the seasons, the cycles of life—these were rhythms that not only defined lives but also ruled them. They were not to be questioned or tampered with, and could hardly be explored. By the nineteenth and twentieth centuries, with the advent of gas and electric light and heat, increasing urbanization, fewer agricultural and more industrial economies, the rhythms of the days and of the seasons were pushed further into the background. Life and work went on after the fall of dusk and throughout the winter months. Time, as a ruler of the body, became lost to view.

There are many rhythms of our lives: the decades of a woman's reproductive cycle and the monthly cycle of her menstrual periods; the twenty-four hours of the day broken into parts by sleep and wakefulness; the seventy beats per minute of our hearts; the hundreds of electrical spikes per second of our nerves. But all of these realms were virtually unexplored until physiologists of the late nineteenth and early twentieth centuries began to investigate the chemistry of the body's tissues. By extracting the active ingredients of tissues and testing their effects on muscle contractions or the electrical activity of nerves, they made a discovery far beyond that of the immediate effect of the material they were testing. What this technology permitted was a rediscovery of time.

After making a kind of soup from glands such as the adrenals—shelling out the inner core, grinding it into salt water, then straining the solid lumps to extract a clear liquid—the physiologists would drip it onto a nerve or a heart or other muscle attached to transducers. And as they watched for the effect, they saw that the nerve generated an electrical impulse, and the heart or muscle beat or contracted. One way or another, they all responded over time. It was impossible to ignore the fact that these extracts had effects that lasted different lengths of time.

If the element of time was obvious in systems like the beating heart, however, it was still opaque in our body's immune defenses. In order to recognize duration of an event, you have to be able to measure some sort of outcome once the system is disturbed. In the case of the immune system, it took many more decades before scientists developed the tools to measure the functions of immune cells and then detect changes in them over time.

The great advances that occurred during the mid-twentieth century in understanding how our immune system works came about largely because of the advent of a handful of new technologies. Central among these was the ability to grow cells in a test tube and then to identify and measure the proteins those cells were making. This confluence of advances in cell biology and protein chemistry in the 1960s allowed scientists to define in ever increasing detail the inner workings of immune cells and the many different functions that these cells could perform.

Instead of stopping at the limits of what is visible under the microscope, scientists could now begin to piece together just what identical-looking cells did in different circumstances. They could add chemicals to such cell cultures—parts of bacteria or plant extracts or foreign proteins called antigens—and then measure whether these chemical soups made the cells grow or divide, specialize or produce proteins. From this they could identify the temporal rhythms of immunity, its immediate and longer-lived responses, the sequence in which these responses occurred, and, most importantly, the kinds of substances that elicit each pattern.

It soon became clear that the immune system responded differently to different stimuli in different time frames. From the moment they are exposed to a foreign protein, immune cells take on new characteristics—characteristics that take time to develop and that evolve as the immune cell matures to its fully specialized function. Although these cells often look the same under a microscope—or may simply look a little bigger, paler, or foamier than before—in fact a beehive of activity is taking place inside the cell that allows it to achieve its full potential. The newly specialized cells accumulate in waves around the foreign material and clear the invader from the body. In this tightly choreographed sequence, timing is critical. Each cell could be functioning perfectly—but any miscue, any shift in timing, and the entire response is put in jeopardy.

By piecing together how immune cells react in culture and what they do in living tissues, immunologists could now reconstruct a moving image. Around the site of a breech in the body's protective armor of skin, blood vessels engorge to bring fluid and red blood cells carrying nutrients and oxygen; platelets and white blood cells to plug the tear and repair the damage are also brought in. The white cells that surge into the area arrive in waves with specialized functions that successively scour the site for foreign matter, at first indiscriminately clearing the debris. The first wave is made up of the neutrophils, with their multilobed nuclei and granules in the cytoplasm. The granules, it turns out, are sacs of enzymes that empty into the inflamed area to break down undigestible debris into more palatable portions. Because these enzymes are irritants in themselves, they contribute to the heat and redness of the inflammation. The next wave, the monocytes or tissue macrophages, large, wavy-edged ghostlike cells with big kidney-shaped nuclei, start to gobble up the foreign protein—"phagocytose" it—by slowly sending out tentacle-like feet that surround the prey. Once engulfed in balloonlike bubbles inside the cell, the invaders are drowned in a sea of enzymes designed to digest them into their tiniest amino-acid parts. These pieces of protein are then recycled back to the cell surface, spit out as it were in a tiny burp, to be stuck onto other proteins embedded in the cell's outer membrane. The bits of invader protein are thus positioned, stuck like flags on a flag-pole, on the host's own macrophage surface. It is in this form that the bits of foreign invader can be recognized by other kinds of white cells, the T lymphocytes. Once they are recognized, the T cells dock onto the invader bit on its protein barb and lock it into a slot on the T-cell surface, a slot that triggers all the rest of the immune system's memory. This protein slot is called the T-cell receptor for antigen. Once the lock clicks, the T cells begin to divide and grow and to spew out all sorts of protein signal molecules—the interleukins. These interleukins call in more waves of immune cells from blood and surrounding tissues, cells that grow, divide, mature, and perform their specialized functions in the complex battle plan. Later still, when immune memory kicks in, new waves of lymphocytes arrive—the B lymphocytes. These turn into those cells that histologists named plasma cells, the oval white blood cells with the skewed nuclei. The plasma cells start churning out antibodies—those bullets that can coat the invaders to help to filter them out of the blood through distant organs such as the spleen.

In all this, there is much movement of white blood cells around the body, from immune organs, such as the spleen and lymph nodes and thymus, to the areas where these cells are needed to fight off invaders—sites of inflammation. Once loaded with debris, the cells, like macrophages, move back again to the spleen or liver, where they are dumped and destroyed, debris and all. The highways along which the immune cells wander are mostly the blood vessels. But there are also the lymphatics—that parallel, ghostly set of vessels that also act as a conduit for these cells. This transparent lacey waterway provides white blood cells with an alternate route of passage around the body, through a slow-moving, lazy set of canals, close to but mostly not connecting with the pulsating, eddying, fast-moving bloodstream. These vessels are filled not with blood but with clear fluid called lymph, from the Latin for "pure, clear water." These thin-walled vessels converge to form several large trunks, which empty their contents into the bloodstream at the confluence of the great veins in the neck. Here, one-way valves prevent the backward flow of blood into the lymphatic system as the lymph spills into the blood draining into the heart.

The entire process, from the arrival of white blood cells at a site to the fluid accumulation (pus formation) and engorgement of the area, constitutes what we recognize as inflammation. The culmination of the sequence is the plugging of the hole and scar formation, which is performed by the fibroblasts. These delicate, spindle-shaped cells don't come from the blood. Rather, they sit quietly within tissues, in spaces between other cells. At the moment when the inflammation has quieted down, when the foreign invader is gone and the debris of battle has been cleared, the fibroblasts start growing and dividing and making a protein called collagen, a tough, interwoven, strandlike protein that is the glue that holds our tissues together. When this happens, and enough fibroblasts and collagen accumulate in one place, a scar is formed.

A myriad of proteins of all sorts work together to smooth this astonishing process of healing. If we could peer into the damaged skin and look at the variety of proteins that are choreographed in the sequence, we would see shapes as many and as varied as the teeming life forms seen in a water droplet by the microscopers of the sixteenth century. There are globular proteins, twisted strands of amino acids rolled up like a messy ball of wool; and flat pleated sheets and tightly braided cords, strong stuff that goes into the mortar to hold our cells and tissues together. There are long, strandlike proteins that snake back and forth and in and out of the cell membrane in seven loops, and other strandlike proteins that can loosely join in pairs with each other or with other unrelated strands.

If you were to plunge through the cell membrane where these proteins glide back and forth along the inner skeleton of the cell, you would find yourself in a silent, clean, energy-efficient world of assembly lines and factories where the building blocks of life are assembled and disassembled with amazing precision and perfect split-second timing. From the distant view of a high-power microscope lens, the cell membrane appears to be a solid, impervious barrier. With more powerful electron microscopes, its fuzzy, spongy outline can be seen. But through the lens of biochemistry, the makeup of that cell membrane comes sharply into focus. It turns out to be made of two layers of globular proteins, each attached to a long fatty tail. These layers line up naturally with the fatty tails facing each other and the water-soluble, oil-repellant heads apart. If you could dissolve yourself first in water and then in fat, like alcohol or ether can do, you could slip silently through this membrane into the cell's internal soup. And there are other ways to traverse the barrier—through pores on the surface, doughnutlike channels that are closed until, like Aladdin's cave, they open at just the right signal. Or you could be quickly engulfed by the tentacle-like pseudopods of phagocytosing cells or carried in on tiny molecular pumps.

It isn't absolutely necessary to cross the membrane to affect the inner workings of the cell, however. The cell surface is not a barren, flat wasteland. Floating in it like icebergs are receptor proteins, whose mountainous tops stick up into the outer fluids through which the cell moves. A small molecule can lodge in a precisely formed cleft—the receptor—of such an iceberg, and perturb it so that signals pass through to the inside of the cell. Other proteins then move along the gossamer net of the cell's inner skeleton to attach themselves to the bottom of the iceberg. In this way, a molecule outside the cell can send a second signal to the cell's inside to activate it's chain of metabolic machinery.

—◄o►—

The Pasteur Institute in Paris was built in the 1870s by Louis Pasteur, at first as an addition to his house. Even today, you can visit the family's quarters by passing from the institute's main hall up through a staircase on the top floor. From his living quarters, Pasteur was able to go to the laboratory at any time of day or night to work on a problem or supervise others in their experiments. Today, in a crypt off the main hall of the institute, you can step through a brass gate and walk down a few stairs to Pasteur's tomb—a mausoleum worthy of nobility, whose design was guided by his devoted wife of close to fifty years. The gold and colored tiles form ornate pictures along the walls and ceiling of the crypt, attesting to Pasteur's many contributions to medicine and society: cows for his discovery of pasteurization, the process for sterilizing milk; dogs for his discovery of the rabies vaccine; sheep for his discovery of the vaccine against the deadly anthrax bacteria (endemic to sheep); winding vines for his studies on purifying wine; and mulberry leaves and moths for his contributions in eradicating the silkworm epidemic. After Pasteur died, many other scientists carried on his legacy, studying infectious diseases and viruses and developing vaccines. Thus they founded the institute, which today carries on the work that helps us understand how the body's immune system copes with invading organisms.

In the basement of the Pasteur Institute, two flights down a set of steps from the ornate marble entrance lobby, is a hallway lined with centrifuges—large square instruments that to the uninitiated look like old washing machines. In fact, these instruments operate on much the same principle as washing machines. Their buckets swing as they gather speed, until spinning at top speed they stick out horizontally, flinging the contents of the tubes within them firmly into a tiny pellet at their tips. You can then safely pour off the liquid and study the solid portion that remains.

It was the development of this method for separating solids from the liquids in which they are suspended that allowed scientists of the 1940s to separate out the tiniest fractions of cells. Through chemical analyses they could then determine the makeup of these particles. The beauty of this method lay in the fact that different-sized particles can be brought down at different speeds. And no matter how many times you repeat this procedure, if you set the speed right, you will reliably isolate the same particles. If you then add a twist—spinning the particles through a thick, syrupy gradient of sugar water—you can separate even more precisely particles of different sizes as well as of different densities. This technique allowed researchers to separate from the rest of the cell soup the tiniest particles, the nucleus, the mitochondria (energy factories), and ribosomes (mysterious, pearl-like structures dotting the endoplasmic reticulum). And that technology eventually led to the discoveries of how genes sitting inside the nucleus control the cell's day-to-day protein production.

It turns out that the ribosomes are like assembly lines for the construction of proteins from amino-acid building blocks. But it is the genes inside the nucleus that direct which blocks will be used and in which order, by sending messenger molecules, called messenger RNAs, out into the cell's cytoplasm. In 1958, three French scientists at the Pasteur Institute—Francis Jacob, Jacques Lucien Monod, and Andre Lwoff—set out in their tiny basement laboratory to figure out how genes control the proteins that are made inside the cell. They found that there are three types of genes that control protein production—one that determines the protein's structure, one that keeps production turned off until the right signal appears, and one that turns on the assembly line. (Jacob, Monod, and Lwoff received the Nobel prize for this discovery in 1965.) It is at this level that genes inside immune cells can initiate the production of all the immune molecules that mediate the cross-talk between immune cells at sites of inflammation.

Zoom back now to wide angle. The cells at the site of an inflammation don't work alone. The many kinds of white blood cells work together in the sticky ooze we call pus. Macrophages work with various lymphocytes, which in turn work together. Some lymphocytes make antibodies; others kill viruses or tumor cells. There are many specialized players in this scene. The complex and intricate process among so many different immune cells requires that the cells have some way of signaling between each other. This allows the right kinds of cells to flood into the injured site, to be available to respond at exactly the right moment.

The signals that allow this communication to occur are also proteins. As advances in chemistry enabled scientists to identify protein products, determine the sequence of the amino acids that make them up, and then produce the proteins in the laboratory, researchers could begin to identify the many invisible signals that immune cells use to communicate with one another.

These molecules were first named by their discoverers according to the function they were found to produce in cultured cells. But in many cases it was found that the same protein had many functions, depending on the test system in which it was studied. A standardized naming system was soon developed for such communicator molecules, called interleukins (from the Latin *inter,* meaning "between," and "leukin" for white blood cell—thus proteins that signal "between white blood cells"). They are now named by number according to their order of discovery: interleukin-1, interleukin-2, and so on past interleukin-20 (IL-1, IL-2, etc.).

Dissolved in the fluid that bathes cells inside the tissues at the site of an inflammation are waves of interleukins, hurled from the waves of cells that infiltrate the area. The first onslaught, led by interleukin-1 and -6, is made by macrophages and has the effect of stimulating the next wave of cells to arrive, the lymphocytes, into making other interleukins, IL-2 and -4. These interleukins make T lymphocytes grow and divide, a process essential in the development of immune memory. Still other interleukins make B lymphocytes grow, divide, and start producing antibodies. In this way, most of these interleukins also act as amplifiers, so that many cells are concentrated in the right place at the time of injury. Other molecules released from these cells can attract still more white blood cells to the area. What starts out as a small brigade quickly becomes a huge army of growing, active cells—churning out proteins and antibodies, chewing up bacteria, and clearing the carcasses of dead and dying cells. In such a maelstrom, perfectly timed signaling between cells is essential.

As in any such escalating war, there must also be an exit strategy— a way of turning off the reaction before these voracious cells turn on the body's own tissues. It was only very recently recognized that a crucial organ involved in orchestrating the exit strategy is the brain. This overdue recognition of the role of the nervous system in regulating immunity stemmed in part from the very technologies that permitted the amazing advances in understanding how the rest of the process works. First amongst those technologies was the ability to grow immune cells in tissue culture dishes and the observation that in this out-of-body setting immune cells worked perfectly well on their own. Exciting as it was, this technology, in the context of twentieth-century reductionist science, led to a thinking among immunologists that the immune system did not require the rest of the body to function. Since these technologies focused the science of immunology increasingly within the cell—on the amazing and delicate molecular machinery that makes these cells function—it also created a mind-set that did not extend past the surface of the cells being studied. So, while such a focus uncovered whole worlds within the cells, it excluded the importance of the worlds beyond—the rest of the body, and especially the brain.

It was the discovery of the between-cell signaling molecules, the interleukins, that gave scientists the tools to prove there could be invisible ways not only for immune cells to signal one another but for the immune system to signal distant organs, including the brain. And it was the further discovery through molecular biology of new ways of synthesizing these molecules in pure form and in large quantities that gave scientists the power to test in living animals the idea that the immune system could signal the brain. Until this was done, no one could imagine the means by which these two systems could communicate from a distance. If one can't imagine it, it seems impossible—and thus not real.

CHAPTER · 4 ·

Putting the Mind and the Body Back Together Again

———◄○►———

At first, only a handful of scientists, convinced that the nervous and immune systems could communicate, attacked the problem, as if beginning at different corners of a jigsaw puzzle. Each had a different motivation for asking their questions, and none realized that they were working on connected parts of the same picture. Neuroanatomists couldn't believe the prevailing dogma that immune organs hung freely in the body, unconnected to the nervous system. These scientists began looking for anatomical evidence that immune organs were wired by nerves. Endocrinologists, trained to accept that glands throughout the body operate in amazing synchrony, had no trouble accepting the idea that such geographically separate glands communicate through soluble molecules, the hormones, which travel through the bloodstream. Why couldn't molecules released from immune cells also act like hormones, affecting organs and cells far away from the site at which they were produced? Then there were immunologists who began to test the effects of chemicals from the nervous system on immune cells. And there were those studying animals highly prone to inflammatory or infectious diseases, who noticed that differences in susceptibility to disease couldn't always be entirely explained by differences in those animals' immune systems. Some set out to find what such other factors could be, and others happened upon them serendipitously. And at the same time as all this was going on, in the 1970s and 1980s, there was another group, at the opposite end of the spectrum from the biologically oriented scientists, who started with the mind and the emotions and asked how these higher brain functions might affect the immune system and disease. No one worked in a complete vacuum, however. The explorations of how these systems talked to each other grew out of fifty years of science that had gone before—the science that had uncovered the workings of the brain and body's stress response.

—◄o►—

At the base of the brain, sitting in its own little seat inside the skull, is a gland that hangs down from a stalk like a cherry. The bony seat reminded anatomists of a Turkish saddle, and so they named it the "sella turcica." And the gland, the pituitary gland, about the size and consistency of a pimento inside an olive, seems part brain, part bodily tissue, so close is it to the back of the nose. When you cut this gland and look at it through a microscope, you can see that it is made of two parts, front and back. The back part is made of long nerve-cell processes that pass through the stalk from their start in cell bodies up in the brain. The front part is suspended in a net of blood vessels. The very different arrangement of these two parts of the gland suggests different functions, though what these functions might be is not evident from the arrangement.

If all you had were your eyes and a set of dissecting instruments, how could you begin to figure out what such a structure might do? One way to begin to assign function is to try to relate abnormalities in anatomy that are found in death to loss of function that existed during life. This is exactly what the physicians of the nineteenth century began to do in autopsy studies of patients who had recently died. In the early part of the twentieth century, the British physician Sir Harvey Cushing noticed a pattern of disease in some of his very obese patients who were different from the rest. They weren't fat all over their bodies—they had spindly arms and legs. And they had accumulations of fat on the backs of their necks very much like the hump of a buffalo. They also had other strange features of disease—spiderlike accumulations of blood vessels underneath their very thin and shiny skin. And many were depressed. They cried, they had trouble sleeping, they were demoralized and discouraged and unable to carry on their daily activities. They felt that nothing they did was good enough. They berated themselves for their incompetence. And some were so despondent that they committed suicide. On autopsy, it turned out that in every case the pituitary gland was swollen with tumor.

But just knowing that an enlarged pituitary gland is associated with such symptoms doesn't tell you what about that gland produces these feelings, or even if it produces them at all. Figuring out the answer to that question requires an entirely different set of technologies. It requires the ability to distill an extract from the swollen gland, to separate each simple component of the complex mixture that results, and to somehow test the function of each of these components. Together, the sorts of chemical analyses that are used in this kind of research comprise the sciences of biochemistry and endocrinology, and were not developed until about the middle of the twentieth century.

What these kinds of studies revealed is that a pituitary gland swollen with tumor makes large quantities of something called adrenocorticotropic hormone, or ACTH. The hormone was given this name because the function that it produces is the stimulation of the outer shell of another set of glands, the adrenals. The adrenals are not located in the brain but down in the belly, perched on top of the kidneys like little hats. Virtually all the clinical features of Cushing's syndrome can be accounted for by the excess of hormones produced by these pituitary tumors. It was the understanding of the hormonal cascade in this extreme case of disease that led endocrinologists later to seek and find similar, though attenuated, patterns of hormonal responses in more normal situations associated with depression. Once it was known that excess cortisol and ACTH were associated with extreme feelings of sadness and anxiety in Cushing's syndrome, it remained only to measure these hormones in the blood of living patients with clinical depression. Such studies led to the discovery that, indeed, in certain forms of depression the concentrations of hormones such as cortisol and ACTH are very high.

On further study, it turned out that the relay race of hormones from pituitary to adrenal did not stop—or rather, begin—there. Later, scientists began to question what part of the brain and what hormone orchestrated this response. It turned out that the part just above the pituitary, called the hypothalamus, started the cascade. A hormone called corticotropin-releasing hormone (CRH) is squirted out into the net of blood vessels surrounding the pituitary gland. The hormone has this name because it stimulates pituitary cells in turn to release their hormone, ACTH. This latter hormone was named for its ability to make the outer shell, or cortex, of the adrenal glands enlarge and pump out their hormones, corticosteroids. Here, then,

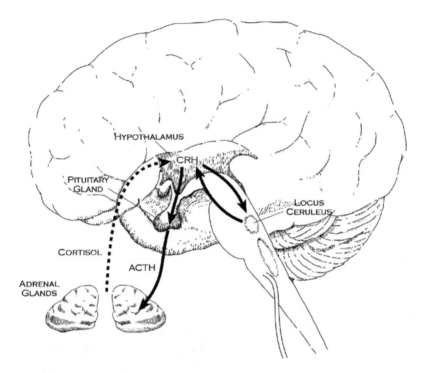

The brain's stress response is made up of many pathways that connect with other parts of the brain and other glands in the body. The command center of this stress response is the hypothalamus, which can be activated by signals coming from the bloodstream or nerves, as well as from other centers within the brain. Once activated, the hypothalamus pours out the molecule CRH (corticotropin-releasing hormone) into the blood vessels surrounding the pituitary gland. This causes the pituitary to produce ACTH (adrenocorticotropic hormone), which stimulates cells in the adrenal glands to make still a third hormone, cortisol. Cortisol acts on many organs and cells throughout the body, including the hypothalamus, where it has the effect of shutting down the production of CRH. This ingenious negative-feedback effect of cortisol is what prevents the stress response from spiraling out of control.

are two glands, half a body apart, connected not by anatomical but by hormonal links. Once the trigger is pulled by any one of many stressors, the cascade is set into motion, and the same sequence of hormones is spilled into the bloodstream. As this sequence was worked out, the inevitable question was asked: If a tumor that produced too much ACTH and cortisol could bring on a characteristic set of symptoms, could similar, but less severe, symptoms be brought on by other situations that caused these hormones to spill out?

—◀◦▶—

On the steep hill along University Street in downtown Montreal, there is an Italianate red brick house at the corner of an intersection across from the McGill University campus gates. From this corner, looking back up the hill, you can see the stone bridge across University Street connecting the Royal Victoria Hospital to the Montreal Neurological Institute, the institute where the neurosurgeon Wilder Penfield and his colleagues carried out their work mapping brain function. If you look up toward the green awning–covered, third-floor solarium of the corner house, your eye will stop just past the ornate stone arch over the door that gives out onto the street. There, carved in a gray stone set into the red brick, is the unmistakable chemical structure of a steroid hormone molecule. This was the home of the Austro-Hungarian–born Canadian physician and physiologist Hans Selye, whose theories on stress turned the scientific community on its edge in the 1950s. It was Selye who introduced the word "stress" into the vocabularies of many different languages, popularizing the concept and proselytizing it worldwide.

When my father knew him, they were both professors of medicine at the Université de Montréal, the French university on the other side of Mount Royal from McGill University, the Montreal Neurological Institute, and Selye's home. In the 1950s, when my sister and I were in grade school, we would walk home every lunchtime, as would my father from the university. From precisely noon to 1:00 P.M., we would sit down and eat the lunch that my mother had prepared in true European fashion. It always included a salad of tomatoes and cucumber; dill and olive oil; and "brinza," the Romanian equivalent of feta cheese. My father concluded with a cup of muddy, foamy Turkish coffee. Sometimes he would bring home from the lab stacks of blue-stained filter papers, still wet and smelling of alcohol. These were paper electrophoreses—a way of separating proteins on paper in an electrical field, which my father had learned in Paris. Like an artist displaying his work, he would proudly line up these filter papers on the credenza opposite my seat at the table. During lunch, if my attention drifted, I would gaze at the blue smudgy lines on the paper— lines that looked as if someone had dipped a too-wet paintbrush in a watercolor box and painted stripes on thick paper. My attention wandered when the conversation between my mother and father sometimes turned in hushed tones his colleagues on the faculty. At such times they would switch from English to their native Romanian so that my sister and I wouldn't understand. But we understood enough to know that there was something darkly mysterious, and definitely different, about a professor whose name was Selye.

The Catholic Université de Montréal was originally founded to train the French Canadian families of the region. But during the late 1940s and early 1950s, at a time when McGill University was much more of a bastion of English culture in Quebec than it is now, the Université de Montréal welcomed into its faculties foreigners and Jews like my father, who had recently escaped war-torn Europe, or like Selye, who had left just before the war.

Selye had left Europe in the decade before World War II and came to McGill University as a young postdoctoral student to study endocrinology. Born in Vienna in 1907, by 1925 he was studying medicine in Vienna and Prague, then among the most prestigious cultural and medical centers of Europe. When Austria began collaborating with the Nazis, Selye moved first to London and then to Canada, where he took up his position as research assistant at McGill University.

From each country, Selye took with him a new language. Fluent in each, he would later lecture to audiences in Europe in their native tongue. My father, trained in Romania before the war as a physician and endocrinologist, followed a similar course, wending through Europe as Selye had and also picking up languages as he went. But he had followed a different route, through the concentration camps of Russia's Dniester River during the war, before reaching Paris and the Pasteur Institute after the war. There he studied the peaceful uses of radiation in medicine with a group of other physicians and scientists inspired by Pierre Joliot Curie. Five years after Selye had taken up his position as director of the Institute of Experimental Medicine at the Université de Montréal in 1945, my father came to Montreal to join my mother in 1950 and found himself in the same faculty of medicine as Selye.

Selye was a slight man with blue eyes, sharp features, and, at the time from which I remember him, white wispy hair. But his personality was larger than life. He thought in grand strokes—his theories were sweeping, and his ego and flamboyance were great. When he first settled in Montreal, he joined the faculty at McGill, where as a young postdoctoral student he had visions of discovering the next new hormone. The department of endocrinology was famous worldwide for a string of such discoveries at this time, when new hormones were being found at a rapid pace in extracts of every conceivable tissue. It was while he was pursuing this dream, and failing, that Selye made the observation that led him to synthesize his theories of the generalized stress response.

In those days, the standard procedure for identifying a new hormone involved making an extract—a kind of soup or infusion—from any organ, often culled from slaughterhouses. The liquid was then injected into a rat, and the effects were observed. Selye was searching for a new female sex hormone, and so he had extracted ovaries and injected the subsequent solution into rats. He noticed that his rats developed enlarged adrenal glands, shrunken thymuses, and stomach ulcers. Enthusiastically believing that he had discovered a new hormone he began to test other organ extracts to determine its specificity. His enthusiasm soon turned to despair when he found that not only extracts of ovaries but also extracts of kidneys, spleen, or any other tissue he tested produced the same effects. In his autobiography he describes the moment of insight when one afternoon, in the depths of his disappointment, his gaze fell on a bottle of formaldehyde in the lab. He mixed a little formaldehyde with salt water and injected a small amount of it into some rats. They, too, developed the same triad of symptoms. He suddenly realized that what he might be observing was a phenomenon more important than the discovery of a new hormone. It might be the body's generalized response to exposure to any sort of contaminated material. He remembered the first patients he had seen as a medical student in Prague, who had all had similar "nonspecific" symptoms of disease. Perhaps this nonspecific response was really a common response to all insults. This was the beginning of what he named the general adaptation syndrome.

Perhaps Selye's theories were too revolutionary and didn't fit into any well-defined niche in endocrinology. Perhaps his ways did not fit with the conservative, restrained culture of the English-language McGill University of the 1940s. Whatever the reason, he left McGill in 1945, moving to the French-language Faculty of Medicine of the Université de Montréal, on the other side of the mountain, to found and direct a new institute. Inspired by French physiologist Claude Bernard's 1865 book *l'Introduction à l'étude de la médecine expérimentale,* Selye named the institute L'Institut de Médecine Expérimentale: the Institute of Experimental Medicine.

The Université de Montréal's classic yellow brick tower, designed by the art deco architect Cormier, protrudes above the hillside of one of Mount Royal's domes, dominating the skyline on a northern approach to the city. In the 1950s you had to climb hundreds of wooden steps, their railings often completely covered by caterpillars, through trillium-filled woods to reach the main traffic circle in front of the tower's soaring windows and main hall. Now you can glide up on a long fluorescent-lit escalator cut through the stone hillside.

The department of medicine was on the ninth floor of the university. From here, you could look out over the city below and watch the sun set over the Laurentian mountains to the north and west. You could walk along the long echoing hallways, whose ochre-tiled walls were pierced periodically by doorways to laboratories. As you passed each doorway, the sweetish, chalky, faintly iodine smell became stronger. These odors emanated from the glass vats of tissue fixatives and stains—mainstays of the biochemist and histologist. If you peered into a lab, you might see rows of such glass troughs filled with colored staining fluids, as well as scaffoldings of glass tubing, microscopes, glass Erlenmeyer flasks and filter paper–lined funnels, all set on satiny slate bench tops.

You could also, if you were a child, run up and down the echoing stairwell at the end of the hallway and carefully position yourself at the top to drop a rolled up ball of paper so that it fell all the way down the middle of the well to the basement. Run up one floor, and the warm urine smell of the animal rooms became intense. Run down two floors, and the fixative smells again permeated the stairwell. And if one of those paper balls you were chasing went accidentally astray, and you sneaked out onto the seventh floor to retrieve it, you would find yourself in Professor Selye's domain. If you were quiet, and could avoid being noticed, you might see the great professor himself stalking down the hallway in his starched labcoat, surrounded by a gaggle of admiring students.

Selye ran his department and institute in efficient and demanding Germanic style. He developed an astonishingly comprehensive bibliographic filing system for scientific papers before the era of computers. He set his graduate students to work in methodical, systematic fashion, gradually increasing their responsibilities in the lab and in writing papers until they were independent. Selye was a stickler for detail and often didn't trust others to carry out experiments—especially in techniques where he excelled, like histology and fine animal surgery. But he could be attentive and generous to his students and they did learn from him, and his more than fifty doctoral students eventually went on to develop their own successful research careers. One of them, Roger Guillemin, received the Nobel prize in medicine in 1977 for his landmark discoveries of hypothalamic hormones.

Selye was a figure to be looked up to and followed—Herr Professor. Stay within the confines of being the admiring student, and he maintained his interest; step beyond the boundaries of student, and his interest waned. Show talent and showmanship approaching his own, and he bristled. These traits derived in part from expectations he had brought with him from Europe; my father had some of them as well. It wasn't so much ego, although there was some of that, but a feeling that once one reached the level of Professor, one should be accorded a degree of reverence appropriate to the station. While this was taken for granted in Europe, it was not the case in North America.

Combined with Selye's ego, flamboyance, and conviction that his outlandish theories would change the world, these traits antagonized other faculty members wherever he went, especially in more conservative academic circles. It might have been different if his theories had been more mainstream. But they weren't. Many at the Université de Montréal were more accepting of his ways than those at McGill had been, perhaps because of the French and continental influence. But they were still a little suspicious and threatened by his new ideas and grandiosity. And ultimately, this very personal, visceral response to the man seeped over into colleagues' ability to accept and embrace his thinking.

Soon Selye began to expand his theory of the general adaptation syndrome to encompass not only nonspecific chemical exposures from the extracts he had injected but all sorts of stressful stimuli. He noticed that the common pattern of illness occurred in animals exposed to any sort of chronic stress, from whatever cause: crowding in the cage, noise, fighting with cagemates. Eventually, if the stress went on long enough, the rats lost weight and seemed to become more susceptible to infection; eventually, if the situation was allowed to persist, they died. On autopsy, these rats, like the ones injected with extracts, had much-enlarged adrenal glands, sometimes two to three times their normal size, as well as enlarged pituitary glands, shrunken thymuses, and stomach ulcers.

For all his independent thinking, Selye was nevertheless influenced by the physiologists who preceded him, Claude Bernard and Walter B. Cannon, whose studies over the previous hundred years had laid out the principle that the body's physiological and biochemical responses are all geared toward maintaining the living organism in a state of balance. In the 1860s, Bernard had spoken of the milieu interne—the body's internal milieu that was maintained in a state of balance. In the 1920s and 1930s, Cannon called this balance within "homeostasis" and proposed that all physiological responses to perturbations that threatened homeostasis were meant to correct the perturbed balance and restore the state of homeostasis.

In his autobiography Selye describes how he found that there was no word in any language to describe accurately general insults that might perturb such a balance. The closest he found was one used by physicists, who used the term "stress" in mechanics to describe a force applied against resistance. Cannon had also used "stress" to apply to general tensions that resulted in perturbation of homeostasis. Selye seized on this term to describe the general insults that could lead to the syndrome he had observed. He aggressively began to promote the concept of biological stress and soon managed to popularize the term in every language and every country where he lectured: "lo stress," "der stress," "el stress," "o stress." He was even able to have the English word "stress" accepted officially as a French term by the Acadè-

Hans Selye, whose theories on stress and its effects on health turned the scientific community on its edge in the 1940s and 1950s, popularized and proselytized the concept worldwide. It was Selye who introduced the word "stress" into the vocabularies of many different languages.

mie Française, the guardians of the purity of the French language. This august body decided, after much heated debate, to create a new word, and give it a masculine gender: "le stress."

Selye proposed that his general adaptation syndrome was caused by an excess of adrenal hormones. He suggested that the adrenal glands were enlarged because they were pushed by the pituitary hormones to pump out extra glucocorticoids in response to chronic stress. In turn, the effects of adrenal hormones caused general weight loss and sickness. But he also noticed that during the early stages of the syndrome the number of some white cells increased in the blood and tissues. These were the neutrophils. Since the ability of these cells and the macrophages to gobble up debris also increased, he predicted that the adrenal glands' corticoid hormones might enhance the body's defenses against invading organisms.

Some say that it was this part of his theories that led many in the academic world to discredit his hypotheses, since it was later found that glucocorticoids were powerful inhibitors of inflammation and immunity. But Selye's observations of the kinds of white blood cells that increased and decreased in the course of stress were accurate. Only some of the interpretations of his observations turned out to be wrong. In part, this misunderstanding arose because in the early days glucocorticoids were used in high doses, essentially as drugs. At such doses they have huge immunosuppressive effects. We now know that at very low doses, however, glucocorticoids can enhance some aspects of immune-cell function. But when Selye was developing his theories, the direction he had predicted for the adrenal hormone's effects—enhancing some parts of the body's defenses—appeared to be wrong. A good part of the scientific community, already resentful and envious of his flamboyant ways, now practically rejected his theories, ignoring even those parts that were correct.

Selye was forceful in expounding his theories not only to other scientists but to the lay public as well. And here he was also charismatic. At a time in science when talking to the lay press was a taboo tantamount to a debasing crime, his proselytizing only further antagonized the most conservative sector of the scientific community. But this same willingness to talk to any audience, his flair for the dramatic, and his willingness to talk on the radio, all enamored him to the popular culture. Eventually, Selye's theories on stress and his photographs began appearing in widely read popular magazines. His picture even appeared in an ad in *Reader's Digest*. After that he was even more ostracized by the scientific community.

It took decades for scientists to come back to Selye's theories on stress and to credit him for the parts that turned out to be right. And it took decades for the science of the immune and nervous systems to mature to the point where it could identify all the molecules and hormones that come into play when the two systems communicate. It wasn't until this communication was well understood that the phenomena that Selye had so precisely and correctly observed in his rats in 1936 (and, as a medical student, in his patients in 1925) could be explained. It took another forty years for immunologists to begin to believe that stress might make you sick in day-to-day circumstances through such hormonal mechanisms.

-◁o▷-

A 30-year-old woman with fine blonde hair sat quietly in a metal chair in the emergency room cubicle. The fluorescent lights made her pale skin seem even paler, almost translucent. She waited expectantly for the report on her urinalysis. I had looked at it under the microscope and seen just a few white blood cells—no evidence of overwhelming infection to complicate what seemed to be a straightforward bladder infection. She had no fever, and her pulse was steady. The intern had wanted to send her home with a prescription for antibiotics but had called me to sign-off on the decision. There was something about her that had made me hesitate. Maybe it was her paleness; maybe it was the fact that the day before she had taken a powerful drug for headaches that she had bought in Hungary. It was a drug that sometimes suppressed the bone marrow's ability to make white blood cells, and so it was not sold in the United States, except by prescription. I decided to keep the woman overnight for observation.

As we were doing the paperwork to admit the patient, the nurse assigned to her cubicle rushed back to the desk to say that the patient's blood pressure had fallen, her pulse had quickened, and the woman complained of feeling faint. As we lifted her onto the gurney, her eyes rolled back and her skin became pasty white, like candle wax. I felt for her carotid pulse—the large artery in the neck, which is the last to go—and couldn't feel it. I reached underneath her breast and only barely felt a thready beat at the apex of her heart. The nurse could not get a reading on the blood pressure cuff. We looked at each other. The nurse grabbed for the intravenous equipment and handed it to me. I called out to her "1 gram Solumedrol and a dopamine drip" as I wiped down the patient's neck, felt two-thirds the way along the collarbone, aimed my large-bore intravenous needle toward the great subclavian vein, and inserted the IV. The nurse hung the drips of Solumedrol, an intravenous form of cortisol, and dopamine as we began to wheel the patient up to the intensive care unit. Almost as soon as we hooked her up to the monitor, we could see the blood pressure numbers begin to climb and the heart rate numbers begin to fall: signs that she was stabilizing.

The woman's diagnosis was not an especially rare one. She had gone into septic shock, a condition in which the blood pressure falls precipitously in the presence of an overwhelming infection with certain types of bacteria, usually those that inhabit the urinary or gastrointestinal tracts. In the 1970s, when this episode took place, it was essential to treat the condition immediately with massive doses of steroids (the Solumedrol), antibiotics, and drugs to stimulate the heart and raise the blood pressure (the dopamine). Without this, most patients would die from the disease. The reason this patient had shown so few signs of infection just before becoming deathly ill was that the headache drug she had taken had suppressed her bone marrow and she had too few white blood cells to produce the usual signs of infection: fever, flushing, and chills. Eventually, with treatment, the patient recovered and went home.

The memory of this case stayed with me, in part because the patient was so young and had very nearly slipped away before my eyes. The experience jolted me into the realization of the tenuousness of the thread on which life hangs. But just as this young woman had so very quickly nearly died, so, too, had her life just as quickly returned, surging back with each drip of intravenous drugs. It came to mind another night ten years later, at midnight, under the pale fluorescent light of the National Institutes of Health (NIH). A student had called me into the laboratory to check some rats that were suddenly and mysteriously dying twelve hours after we had treated them with a new experimental drug. It was a drug that blocked the effects of the brain chemical serotonin. A couple of years earlier, at Washington University in St. Louis, I had found that when added in tissue culture to macrophages—those nonspecific garbage-collector cells of the immune system—the drug stopped the cells from gobbling up small plastic beads, or anything else for that matter, that we added to the dish. At the time I thought we had discovered a new anti-inflammatory drug, a drug that could perhaps be used to treat arthritis and other autoimmune/inflammatory diseases.

There are many diseases that fall within this category of autoimmune/inflammatory diseases, illnesses as familiar as rheumatoid arthritis, systemic lupus erythematosus (SLE, or lupus), thyroiditis, multiple sclerosis, fibromyalgia, and chronic fatigue syndrome, and other rarer types such as scleroderma and Sjogren's syndrome. The thing that all these diseases have in common is that the immune system turns on itself. Those immune cells usually vigilant against foreign invaders begin to devour and destroy the body's own tissues. They do not do this indiscriminately, but attack specific organs or tissues. In the case of thyroiditis, the thyroid gland is attacked; in the case of multiple sclerosis, the myelin sheaths that wrap around nerves are targeted; and in rheumatoid arthritis, it is the lining tissue of the joints, the synovium, that is destroyed. (Not all arthritis falls into this category of illness—some is caused by wear and tear on joints, not inflammation; however, it is the inflammatory type that can eventually lead to crippling.) Often in such illnesses the immune system turns against the blood vessels, so that any organ in the body can be affected, since blood vessels coursing through them become inflamed and scarred. Although these illnesses can be mild, involving just a little pain and swelling at the areas of inflammation, they can also be life-threatening if the blood vessels, heart, lungs, or kidneys are affected. In the course of disease, waves of immune cells move into a tissue or organ and, in the process, destroy its normal architecture. Eventually scar tissue forms. If the destruction is severe enough, there can be permanent damage and crippling, as in rheumatoid arthritis, or loss of memory, paralysis, or blindness as in multiple sclerosis. Autoimmune diseases that attack the lungs can lead to lung scarring and difficulty in breathing.

Although scientists do not fully understand the cause of these diseases, it is thought that there is generally some environmental trigger—some unknown protein, chemical, bacteria or virus—that starts the process in a susceptible host. Among the factors that make some hosts susceptible to such diseases is an inherited tendency for their immune cells to respond more vigorously to certain foreign chemicals, or antigens. This has to do with proteins that stick out on the surface of white blood cells—those little "flag-pole" proteins to which foreign material, digested within the cell and spit back out, sticks. In part, it is the tightness of the fit of the bit of foreign protein on the flag-pole (called MHC or HLA antigens) that determines how strongly the immune system will react to the foreign invader and whether the immune system will then attack similar proteins in the body's own tissues. If such foreign antigens resemble the proteins in certain tissues or organs, these, too, will be attacked. In the mid-1980s, when I was working on what I though might be a new arthritis drug, much of the research in this area was directed at trying to determine the nature of the foreign-antigen–immune-cell interactions. Not much was directed at figuring out what other factors might play a role in making a host more or less susceptible to developing arthritis.

When I first arrived at the NIH, I wanted to find out if the drug I was testing in cells in culture could really be used to treat arthritis. In order to do this, I convinced Ronald Wilder, an arthritis expert there, to work with me to test the drug in the arthritis-susceptible rats he was studying. Ron and his collaborator, immunologist Sharon Wahl, had found that certain inbred strains of rats developed arthritis when exposed to pieces of streptococcal bacteria. But another rat strain, matched with practically identical MHC proteins on white blood cell surfaces, was resistant to developing arthritis when exposed to the same strep bacteria. So it seemed that something more than just the immune system was needed to predispose or to protect these rats from getting arthritis.

The arthritis-susceptible strain of rats, called Lewis rats, had been bred over many generations by veterinarians seeking to develop a purebred rat strain to be used for testing chemicals that might cause such inflammatory diseases as arthritis, multiple sclerosis, thyroiditis, or adrenalitis. At the same time, veterinarians had bred a related strain of rats for their resistance to inflammatory disease. These were called Fischer rats. Just as dogs or horses can be bred over generations to emphasize certain traits—length and color of fur, racing speed, personality, and shape—so can laboratory rats be bred for other traits. Immunologists interested in studying which components of bacteria might cause inflammatory diseases such as arthritis start their experiments by ordering one or two of these purebred rat strains. By controlling their experiments in this way, researchers know that the degree of inflammation they observe in response to a test agent will be relatively constant, not muddied by variations arising from differing genetically determined "host responses." Therefore, the characteristics of the substance that precipitates disease can be studied without confusion.

The problem with this approach is that no one bothered to ask why it was that these purebred rats were either highly resistant or highly susceptible to inflammation. Everyone doing this research just knew that if you wanted to study arthritis or multiple sclerosis or thyroiditis, you had to start with Lewis rats. It was so much a part of the experiment that it was taken for granted.

That morning, my student, Mike Wechsler, and I had injected the rats—both the arthritis-susceptible Lewis and arthritis-resistant Fischer strains—with either the strep bacteria alone, the strep and the antiserotonin drug, or the drug alone. The second dose of drug was due at 10:00 P.M. But Mike, arriving late in the lab after a game of basketball, called me at home at midnight to say simply "Come quickly—all the rats are dying." I arrived in the animal room shortly thereafter to find that, in fact, not all the rats were dying, just those that were usually disease-resistant, the Fischer strain, and those that had received both the drug and strep cell walls.

As we walked along the long, deserted corridor at midnight, Mike asked how a drug that worked so well to stop inflammation in cells could cause these rats to die so quickly. That was when I remembered the young woman in the emergency room. The way the rats succumbed so quickly after exposure to bacteria resembled the young woman's septic shock. But why would *these* rats die—the ones usually so resistant to disease? And then I remembered how it took huge doses of steroids to save the woman. Maybe, I thought, our own bodies sometimes need to give themselves a shot of steroids. Right when we need it most, when we're exposed to large quantities of bacteria, our own adrenal glands give us a steroid kick. That's when our immune system would be massively activated, pouring out huge quantities of interleukins and other inflammatory chemicals to fight off the bacteria. And while these chemicals and the white blood cells they activate help kill bacteria, in too large a dose they can also be harmful. Just as these chemicals can damage bacteria, and so kill them, in massive quantities released all at once during septic shock, they can damage cells lining blood vessels, make them leaky and so lower blood pressure. If for some reason at that moment our bodies couldn't stop all these mediators from pouring out, then we would go into septic shock. One reason we might not be able to shut off the immune response, I thought, would be if our adrenal glands couldn't make enough steroids—those hormones that turn off immune cells. So, if the drug we were giving to the rats blocked their ability to make adrenal steroids, it could undercut their own protective mechanisms—and explain their shock. In fact, I remembered, the literature on the drug said that it did just that—it crossed into the brain, blocked the hormonal stress response there, and with it blocked the adrenals' steroid push.

There was one other piece of the puzzle that also now fit. How did the brain's stress response know that the strep cell walls were there, and so when to pump out hormones? At this time, most immunologists and neurobiologists did not believe that the immune system could communicate with the brain. But signals from the immune system must be able to turn on the brain's stress response, I reasoned; otherwise it would be impossible to time the steroids correctly. If the adrenals poured out steroids before the exposure to infection, they would decrease the immune cells' ability to fight. The body's shot of steroids had to come just at the right time, after its massive release of immune cytokines. So there had to be some signal mechanism by which the immune system let the brain know when to start its pulse.

That week I had seen a pre-print of a new study by neuroscientist Candace Pert showing that there were receptors in the brain for the immune signal molecule, interleukin-1. Candace was the chief of the section in which I worked at National Institutes of Mental Health (NIMH). She was one of the neuroscientists who, 15 years before, had discovered that the brain had receptors for morphinelike molecules. She was one of the first to believe, and then show, that the brain also had receptors for immune molecules. It turned out that some of these receptors were near the hypothalamus. So, if IL-1 could bind to brain tissue, it must be doing something there, I thought. Maybe one of the things it did included turning on the hypothalamus and the hormonal stress response. All that remained was to test the hypothesis, by measuring these rats' stress hormone responses to an injection of strep cell walls. If we were right, the arthritis-susceptible strain would have low stress hormone responses, since that would make them unable to shut off inflammation, and the arthritis-resistant strain's response would be high, since that would shut off inflammation before it got out of control. The stress hormone that is easiest to measure in blood is corticosterone, the rat equivalent of the human steroid hormone, cortisol.

The next morning we tested five rats of each strain by injecting strep cell walls and collecting blood an hour later. We sent the blood to a local lab for measurement of corticosterone and waited anxiously for the results. As the lab technician dictated the results over the phone, I could see the pattern that was emerging—a pattern that held true every time we repeated these experiments and others like it. We were right. The arthritis-susceptible rats had hardly any corticosterone response, while their disease-resistant cousins' responses were sky-high. Our experimental drug did not turn out to be the next great cure for arthritis. But we had found a clue as to why some rats, and maybe people, were susceptible while others were resistant to arthritis and other inflammatory diseases.

As encouraging as these results were, they amounted to just one experiment. Where did we go from here? Both Ron and I had trained as rheumatologists, not endocrinologists. We knew nothing about the brain or the body's stress hormonal responses. I had no idea how to prove that the reason for these rats' low corticosterone response was in their brains. I showed the hastily plotted graph to the psychiatrist and neuroscientist Steve Paul, who was the chief of the branch in which I worked. He looked quizzically at the penciled dots and said, "You've got something interesting here, but you've got to prove it—show that this is coming from the brain and not the adrenal glands. You need to measure corticotropin-releasing hormone in the hypothalamus directly." That seemed an impossible task. I'd never done it and didn't know it could be done. I said as much to Steve, who replied simply, "Why don't you just ask two of the world's experts? They're both in this building and one of them is just 300 yards down the hall at the other end of this corridor."

The experts Steve was referring to were Phil Gold, a psychiatrist and endocrinologist at NIMH who studied the role of the stress hormones in depression, and his longtime collaborator, George Chrousos, an endocrinologist at the National Institute of Child Health and Disease at NIH who studied glucocorticoids and the hormonal stress response. I had seen Phil wandering the hallways, a short, absent-minded-professor-looking type, with conductor-long curly gray hair. George, I didn't know at all. So I did a Medline search on their publications and found they'd published hundreds of articles on the hormonal stress response in Cushing's syndrome and in depression. As it turned out, they were presenting an afternoon seminar at the hospital's main lecture hall.

Pushing open the door at the back of the dark amphitheater, I could see there were no empty seats left at all. The speaker, a sturdy man with salt-and-pepper graying hair, who seemed to be about my age, then in his late thirties, was outlining the endocrine basis of the stress response in clear and organized detail. I concluded, because of his worldly Greek-American accent, that this must be Chrousos. Gold spoke next and connected the stress hormone response that George had described earlier to psychiatric illnesses. He described their work showing that some patients with depression had a hypothalamic stress hormone response stuck in the "on" position, while in others the stress hormone response was greatly reduced, indicating that imbalances in one direction or the other could trigger depression.

I had never thought of psychiatry in this way. At the time, most physicians thought of psychiatric illnesses as mysterious groupings of symptoms—exaggerated emotional responses without a known cause, and without a cure. Although I knew that imbalances of some brain chemicals, such as serotonin, were involved, it never occurred to me that these illnesses of the emotions, actually illnesses of the brain, were also imbalances of the many brain hormones that work together to produce our feelings. That specific psychiatric symptoms could be so logically connected to equally specific brain hormone changes was amazing. If this was so, it meant that specific drugs could be used to right the imbalances. And if this brain hormone that played such an important role in mood turned out also to be the molecule that had gone awry in my rats, it could be the brain, not the pituitary or adrenals, that controlled arthritis. If this was so, perhaps it could also explain why some arthritis patients have a tendency to suffer from depression. As I stood there listening, I immediately decided that I needed both George's and Phil's input and experience to help design the next set of studies to prove my theory.

The next day I arrived at Phil's office, graphs in hand, and stood at his door waiting for him to finish speaking to a student. The afternoon light was waning, and his office was dim. As he ushered the student out, he asked how he could help me. I explained my question and showed him the data, and as I did so his eyes widened and he became excited. He started speaking quickly. Of course, he and George would help. They could help design the experiments to show the adrenal and pituitary glands' responses in detail over time in both rat strains, and in outbred rats as well. And they could help us measure CRH in the hypothalamus, too. They had two postdoctoral students from Italy, Aldo Calogero and Renato Bernardini, working for them who were whizzes at taking out rat hypothalami and growing them in a tissue culture dish. They could see whether the hypothalami secreted CRH in culture, and they could measure CRH in the hypothalami directly, too. If we would give them some interleukin-1, and also some rats of both strains, they could do the rest. All I had to do was help out with the experiments and, in the process, learn to do them, too.

Aldo was a stocky Italian, friendly and willing to teach. He described what he was doing as he carefully rolled the walnut-shaped rat brain over in one gloved hand, so that we were looking at its underside. Then in four quick snips with a fine-curved surgical scissor, he deftly lifted out a tiny square piece of brain tissue with his other hand. As he snipped, and without looking away from the delicate task, he described the coordinates where the cuts should be made, equidistant around a small dimple at the base of the brain—an unvarying landmark that told you by surface markings that you were near the hypothalamus. This dimple on the underside of the brain, in fact, was where the pituitary stock inserted: the part of the brain that formed the roof of the sella turcica.

"Don't cut deeper than a couple of millimeters, or you'll be taking more than hypothalamus," Aldo instructed. "And don't squeeze too hard when you cut, do it quickly and cleanly; otherwise you will damage the brain tissue and it will swell." He gently placed the tiny square into a tissue culture well, half filled with pink, nutrient-containing culture fluid—one of twelve identical penny-sized circular wells in a twelve-well plastic tissue culture dish. He continued, "If it swells too much then the nutrients in the culture fluid won't get to the core to keep the tissue alive, and any CRH you measure in the fluid may just be leaking out of dying tissue. If you're careful, and..."— here he paused as he carefully picked up the tissue culture dish, now filled with twelve tiny squares of brain in their pink culture medium. He gingerly carried the tray to the warm humidified incubator a few steps away, first opened the outer insulated metal door, then unlatched the inner clear plastic door covered with beads of condensed water vapor, and gently placed the plastic dish on the top metal rack. For the first time, he looked back at me as he quickly closed both doors, and finished his sentence. "If you're careful, and immediately put the tissue at 37° centigrade, in a 5 percent carbon dioxide atmosphere, it can live perfectly well overnight. But we're going to remove it in twenty minutes to wash the old medium out and put in fresh medium—we do that three times and then let the tissue stabilize before we add the drug we're testing. In your case, we'll add the IL-1. Did you bring it with you?" I handed him the plastic tube containing IL-1 dissolved in salt water.

The next day, after we had added different concentrations of IL-1 to each well with a plastic-tipped pipette, I helped Aldo collect the fluid surrounding each piece of brain tissue, so we could measure how much CRH each hypothalamus had put out. This would be done by adding antibodies tagged with a radioactive label. The antibody specifically attached to CRH. The more CRH in the culture fluid, the more labeled antibodies added to the fluid would bind to CRH, and the higher the radioactive counts would be. When the experiment was done, we watched the paper printout from the radioactive counter to see whether there was a difference in the counts in fluid surrounding Lewis or Fischer hypothalami. As each number clicked on the ticker-tape-like printer, we could see that a pattern was emerging: there were groups of low numbers and groups of high. And later, when we analyzed the data, it turned out that it was the fluid from the Lewis hypothalami whose counts were low, and from the Fischer hypothalami whose counts were high. It was just what we'd predicted and hoped for. Though stimulated by IL-1, hypothalami from (disease-susceptible) Lewis rats, even when taken out and grown in culture, put out too little CRH—no more than the unstimulated levels secreted by hypothalami not exposed to interleukin-1. Fischer rats' hypothalami made too much. This meant that the low adrenocorticotropic hormone (ACTH) and corticosterone we'd seen in the blood of Lewis rats resulted from low secretion of CRH from the hypothalamus. And it meant that the high hormone levels we'd seen in Fischer rats came from their high hypothalamic secretion of CRH.

We couldn't stop there. Just because the hypothalami secreted too little CRH didn't mean they made too little. It could be that they made it but weren't able to spit it out. So we did two more experiments. In the first, Aldo measured CRH inside the hypothalamic chunks, and found that Lewis' pieces contained less CRH than did the Fischer pieces. In the second experiment, Scott Young, another scientist at NIMH, measured whether the CRH gene itself worked. And here again, the Lewis hypothalami made less of the CRH gene's product, CRH messenger RNA, than did Fischer hypothalami. So, from top to bottom, from the hypothalamus to the pituitary to the adrenal glands—the so-called hypothalamic–pituitary–adrenal axis— Lewis rats made less hormone and Fischer rats made more. This, then, must be the reason that one strain of rats was more predisposed to develop all sorts of inflammatory diseases and the other was relatively resistant to them all. Finally, we asked one more question, then, in these high-CRH/low-CRH responsive rats. Since CRH controls not only the pituitary-adrenal hormone burst, but also stress-related behaviors, would these rats also *behave* differently in response to stress? Together with the behaviorists John "Jace" Glowa, then at NIMH, and Mark Geyer at the University of California at San Diego, we tested the rats' behaviors and found that indeed the high-CRH

Fischer rats showed high-stress behaviors while their low-CRH Lewis cousins remained relaxed and unperturbed. We had come full circle. If these traits were associated in Lewis and Fischer rats for hormonal reasons, it could also explain why depression often accompanies arthritis in humans. Moreover, this fit with Gold and Chrousos's findings relating CRH imbalances to depression.

But simple association does not prove cause. We did still two more experiments. We gave Lewis rats very low doses of steroid hormone to see if we could prevent them from developing arthritis when exposed to strep cell walls. And we gave Fischer rats drugs that blocked the glucocorticoid receptor to see if their disease would worsen. In both these situations, we again found just what we had predicted. The low-dose steroids blocked disease in Lewis rats, and the glucocorticoid-blocking drugs made illness worse in the Fischers. We were able to change inflammatory disease susceptibility by interfering with or supplementing the hypothalamic–pituitary–adrenal axis. This had to mean that the brain's hormonal stress response played a very important role in modulating susceptibility and resistance to inflammatory disease.

CHAPTER · 5 ·

It's a Two-Way Street

The Immune System Talks to the Brain and the Brain Talks Back

————◄○►————

*I*t was the Swiss immunologist Hugo Besedovsky who first proposed, in 1975, that the natural burst of glucocorticoids released from the adrenal glands during the early phase of an immune challenge actually changed the way the immune system reacted to a second and different challenge. Working at the Schweizerisches Forschungsinstitut Medizinische Abteilung, a research institute in Davos Platz, Switzerland, Besedovsky postulated this because of his observations first in rats exposed to red blood cells from a different species, and then in mice infected with a virus. In both these circumstances, the animals reacted to the foreign substance first by producing antibodies, and then with a rise in blood corticosterone.

Besedovsky had immunized rats by injecting them with the red blood cells of a horse, so that the rats would develop antibodies to these cells. At the same time, he measured levels of corticosterone in the rats' blood and found that it increased during the first six days after immunization. If at that point—at the peak of the corticosterone response—he then exposed the rats to a second and different immunization, this time with sheep red blood cells, the rats were unable to make new antibodies to the sheep red blood cells. But if he repeated the same experiment in rats whose adrenals had been removed, they were perfectly able to make antibodies to the new challenge. So Besedovsky concluded that the reason rats ordinarily couldn't make antibodies to a second immune challenge received shortly after a previous one was that the corticosterone from the adrenals suppressed the immune response. This finding fits perfectly with the course of infectious illnesses in our daily lives: if we get the flu or a cold, it is a couple of days after the viral infection has begun, just when we have begun to get better, that a secondary bacterial infection, such as a sinus or ear infection, usually hits.

In retrospect, this observation also opened the door to the idea that the immune and endocrine systems mount a coordinated response to an immune threat, but at the time it was not hailed as a groundbreaking discovery. In the 1970s, many papers were being published on the effects of glucocorticoids on immune responses of all sorts. Scientists such as Tony Fauci, now known for his role in leading U.S. government AIDS research, published entire books on the effects of glucocorticoids on the functions of immune cells in tissue culture and in the whole living organism. In that setting, Besedovsky's study was just one of many papers exploring yet another condition in which these drugs affected the immune response. And that is just how the hormones were thought of: as drugs. In part, this was because it had been known since the 1940s that they worked in druglike doses to suppress all sorts of inflammation. Edward Kendall, Tadeus Reichstein, and Philip Hench even received the Nobel prize for this discovery in 1950. (This was the work that put the last nail in the coffin for Selye's prediction that adrenal secretions could enhance immune function.) Thus, by the end of the 1970s, immunologists were expressing a "so what" attitude to such studies testing the effects of corticosteroids on immune function. The prevalence of this attitude only increased in the 1980s with the sudden revolution of molecular biology. When this new technology burst onto the scene, studies that focused on such mundane topics as hormonal effects on immunity were largely ignored.

But Besedovsky persisted, following through with another set of experiments that took him one step closer to defining what happens in natural situations. He asked whether the same sort of corticosterone rise and suppression of antibody formation occurs in the context of a viral infection. In 1985 Besedovsky, his wife Adriana del Rey, and their colleague Ernst Sorkin published a paper in *The Jour-*

nal of Immunology examining the effects of living virus on hormonal responses. The paper showed that mice inoculated with a virus called Newcastle Disease Virus developed a large increase in the hormones corticosterone and ACTH in their blood.

In the mid-1980s, another group of Swiss scientists led by Jean-Michel Dayer from Geneva had begun to work on a similar question, but Dayer turned the question around and asked whether brain cells could make immune molecules. He and his group first showed that IL-1 activity could be detected in the nutrient fluid surrounding brain tumor cells grown in culture. They then set out to test whether IL-1 could be made in the brain of animals exposed to infectious products. They also wondered whether IL-1 was possibly the brain chemical that caused fever (since fever, a part of the immune response, would appear to be centrally controlled and therefore to originate in the brain). To answer these questions, Dayer and his colleagues injected bits of bacteria into the bellies of mice in order to mimic infection. These bacterial parts are called lipopolysaccharides, or LPS (for their chemical constituents: "lipo" meaning "fat" and "polysaccharide" referring to a long chain of sugar molecules). LPS is a fatty part of bacterial cell walls, much like the strep cell walls we later used to induce arthritis in our rats. It is the chemically active component of bacterially contaminated "dirt" and acts as an extremely powerful stimulus to the human immune system, even in the absence of living bacteria.

In retrospect, the Swiss scientists were doing the same thing in this experiment that Selye had done, unwittingly, back in 1936. Although he had no way of knowing it, there was very likely bacterial contamination in all of Selye's hormone extracts, and it was probably these pieces of dead bacteria that constituted the contaminating "dirt," thus producing his general adaptation syndrome. At the time, immunologists had not yet recognized that such bacterial dirt can induce immune cells to make large amounts of their hormonelike molecules, later called the interleukins. In retrospect, Selye really had discovered the new hormone of which he had dreamed. But it wasn't a sex hormone or a stress hormone in the usual sense; instead it was probably interleukin-1 and other cytokines. The tragedy is that he died just before such immune hormones were discovered.

By the time Dayer and his colleagues, Elisabeth Weber and Adriano Fontana, performed their experiments, it was well known that immune cells could make signal molecules, such as interleukin-1, after they were exposed to bacterial parts. So why not test other tissues, such as the brain, to see if it, too, could make such molecules? In those days, however, there was no easy chemical way of identifying IL-1. The best way to detect its presence was to see if the material you thought contained it made lymphocytes divide. In order to do this, you had to extract the tissue, make a kind of soup from it, and strain the liquid so that it could be added to cells in culture. The Swiss scientists did just this, making a fine suspension of the brain tissue collected from the mice injected with LPS as well as from brain tissue of mice that had been injected only with a small amount of salt water. They then placed these samples in a very fast centrifuge to bring even the smallest particles and parts of cells out of suspension. They poured off the clear liquid brain extracts from the top of the centrifuge tubes, and added some to a culture dish containing the white blood cells, lymphocytes.

As expected, it turned out that the clear liquid extracted from the brains of mice that had been injected with bacterial lipopolysaccharide made the lymphocytes divide, and the liquid extracted from the brains of saline-treated mice did not. And when they injected that same liquid extracted from brains of LPS-injected mice into other mice, those mice also developed fever. So, a factor could be found in the brains of mice exposed to parts of bacteria (LPS) that stimulated lymphocytes to grow. Usually this factor came from those big garbage-collector cells, the macrophages. But now it seemed some cells in brain tissue were making it as well. And the previous tissue culture studies with tumor cells suggested that maybe it was the astrocytes.

It turns out that many kinds of brain cells make IL-1: small round cells called "microglia," brain cells that are related to macrophages, as well as large brain cells, the astrocytes, named for their starlike shape. These are the same cells that are painted on the ceiling of Wilder Penfield's Montreal Neurological Institute. In the brain, they form a kind of scaffolding for neurons, and we know now that they produce all sorts of compounds to support the growth and life of nerve cells. But in the early 1980s, their function was still a blur, and the discovery that IL-1, a growth and differentiation factor for immune cells, could be detected in the brain and in astrocytes provided one of the first glimpses that such molecules might play an important role in maintaining the health not only of immune cells but of nerve cells as well.

When I came across Dayer's paper in 1984 while finishing my postdoctoral studies, I could hardly believe what I was reading: interleukin-1 activity could be found in the brain. The name "interleukin" had only just been given to various proteins that immunologists were isolating out of the culture medium in which their immune cells grew. These factors were first defined according to their effects on the growth and function of other immune cells, and they were given names that reflected their functional effects. Then, as their chemical structures became known, it was recognized that a single protein in this family could have multiple effects, and so the common term "interleukin" was settled on to identify them. The idea that such proteins could be manufactured in the brain, when they were so clearly made by immune cells in tissue culture, unattached to any specific body part, went headlong against prevailing dogma. In fact, many immunologists argued that these experiments didn't prove that the IL-1 being measured came from the brain. After all, they argued, the injected bacterial products caused inflammation, and inflammatory cells such as macrophages could squeeze past blood vessel walls into the brain. The IL-1 activity could simply be coming from blood and tissue macrophages. Or there could have been some small, undetectable impurities in the preparations, producing the observed effect. Convincing research that IL-1 was indeed produced in the brain was needed.

My supervisor at the time, Dr. Charlie Parker, head of the allergy and immunology division at Washington University in St. Louis, had said to me when I reviewed the paper at a seminar: "Now there's a project for you. If you really want to study something in the brain, find out if interleukin-1 is actually there." At the time I was finding that a drug used to block the neurotransmitter serotonin, the one I eventually tested in the Lewis and Fischer rats, also blocked the activation of macrophages in the tissue culture dish. This was exactly what all the other cellular immunologists at the time were doing: adding drugs to immune cells in culture and watching to see what happened. I was so excited by the possibilities of where the discovery might lead, and the thought that I might have a new arthritis cure, that I stayed my course and decided not to follow Parker's advice. But that small piece of information, that interleukin-1 activity might be found in the brain, stuck with me, to contribute later to my understanding of what happened to those rats at midnight.

At that time, because the stimuli used by both Dayer and Besedovsky—parts of bacteria or whole viruses—were not pure, these early studies showing on the one hand that immune signals could stimulate the brain's stress hormone cascade, and on the other that the brain itself could make immune molecules, were viewed with skepticism by the immunology community. There could be many reasons, it was felt, that the brain responded to such complex materials, made up, as they were, of many different chemical constituents. The brain didn't have to be reacting to an immune signal.

It took the development of two new technologies to prove definitively that pure immune signals stimulate the brain's stress response. And it took yet another advance to prove that brain cells could also make such pure immune molecules. The first technology made it possible to produce huge quantities of pure IL-1, amounts large enough to inject into animals as large as a mouse or rat. And since it was also necessary to show that it was the brain being stimulated, not just organs in the periphery, a second technology for directly measuring the brain's stress hormone, CRH, was also needed. Until that time it had only been possible to check blood levels of corticosteroids and ACTH, but measuring CRH directly could prove that it was the brain that responded to IL-1, not just the adrenal or pituitary glands. It was recombinant technology—popularly known as genetic engineering—and peptide chemistry that allowed all of these proofs. When these technologies arrived, the experiments proving beyond any doubt that the immune molecule interleukin-1 did indeed stimulate the brain and that the brain did produce interleukin-1 were performed. They were all published in the journal *Science*.

—◄o►—

In the mid-1980s a young Dutch scientist named Frank Berkenbosch joined Besedovsky and his wife Adriana del Rey. Berkenbosch followed the same procedure that Besedovsky and del Rey had followed when they tested the Newcastle Disease Virus in mice. And he followed the same procedures they used the year before when they first obtained pure recombinant IL-1 from Charles Dinarello, the scientist who first cloned the molecule. In that study, Besedovsky, del Rey, and Dinarello had injected the IL-1 into the bellies of mice and measured levels of corticosterone and ACTH in the blood. They found that within one hour after IL-1 injection, corticosterone and ACTH levels increased in the blood of mice treated with IL-1, but not in those treated with saline alone.

When Berkenbosch arrived in the lab, Besedovsky and del Rey needed to prove that the adrenal and pituitary glands were being driven by the brain's stress hormone, CRH, to make the plasma corticosterone and ACTH rise. So, Berkenbosch repeated the experiment again, injecting the recombinant IL-1 into the bellies of mice, but this time, in addition to measuring blood hormone levels, he also measured CRH expression within the brain. This would tell him whether the rise in blood hormones he was measuring came from the brain or only from glands outside the brain—the pituitary and adrenal glands. To do this, he used an antibody to CRH, tagged with a fluorescent dye. If the fluorescent-labeled antibody found CRH in a brain slice, it would bind to it, and the CRH-containing cells would shine green against a pitch-black background under the fluorescent microscope.

As he sat in the darkened room and peered down the objective of the fluorescent microscope, Berkenbosch saw that the cells from mice that had received IL-1 shone bright. In that same brain area in the mice that had received only saline, the cells were just pale ghosts. This meant that the hypothalamic cells of mice that had received IL-1 made much more CRH than those of mice not treated with it. Furthermore, it took three hours after the injection for the brightly shining cells to dim. They did dim, however, suggesting that the brain's response to IL-1 was a transient process, but one that lasted long enough to trigger many effects. It lasted long enough to turn on the pituitary and the adrenal glands to pump out their stress hormones, and maybe even long enough to allow these hormones to begin to turn down the immune response. These findings became the core of the paper that Besedovsky's group published in *Science*.

The beauty of these studies was that the two technologies that made them possible—one allowing the testing of the brain's stress hormone, CRH, and the other allowing production of large quantities of IL-1—had ripened simultaneously. So that now, for the first time, the separate disciplines of endocrinology and immunology were ready to fuse their methods, to prove a connection between the nervous and immune systems. The group of scientists who had made possible the measurement of CRH—by isolating it, identifying its amino-acid sequence, then making enough of it to inject into rabbits so as to develop antibodies to use on those CRH-containing brain cells—was a group of scientists at the Salk Institute in La Jolla, California, led by an endocrinologist named Wylie Vale.

◄○►

Through the glass-walled labs of the Salk Institute you can see hang gliders fly their red and yellow and blue contraptions off a nearby cliff over the Pacific Ocean. It is here that the Swiss endocrinologist Catherine Rivier came when she left her home of Lausanne overlooking Lake Geneva. She and her husband, Jean Rivier, worked as a team—he a peptide chemist, she an endocrinologist/physiologist whose expertise was testing effects of peptide hormones on the brain. They had joined Wylie Vale at the Clayton Foundation Laboratories for Peptide Biology, which he directed. Together, they had discovered the structure and the functions of the stress hormone CRH. And together they had developed other peptides—drugs that mimicked or blocked CRH's actions on the brain and hypothalamus. And with large quantities of these peptides and the drugs or antibodies that blocked them, they and their colleagues were able to determine all the biological effects of CRH on the brain and its stress response.

Down in the darkened laboratory surgical suite, Rivier focused the bright surgical light on the tiny area on which she was operating: 1 square centimeter at the top of a rat's head. Here she would soon insert a needle to guide a tiny plastic tube. The rat, fully anesthetized, felt no pain. Its head was held immobile in a metal crown, marked with a fine-numbered grid in three directions, like the coordinates of a three-dimensional graph. The grid held the needle's base, and by spinning knobs to set coordinates in each plane, the angle of the needle could be shifted. The goal was to set the needle at just the right angle so that, when inserted in the skull, in one deft move, it would reach the tiny spot of brain where Rivier aimed it, and go no further. This spot was the hypothalamus, deep inside the brain, about half an inch from the surface of the skull. Just as Penfield and Cone had worked out the coordinates on the surface of the human skull that helped to guide the neurosurgeon's electrodes deep into brain without damaging surrounding tissues, neuroanatomists expert in the rat had done the same. They found the surface distances of the angle and the depth a needle should go in order to reach a desired goal.

Once the needle was in place, and a plastic tube inserted in the spot, Rivier could inject whatever drug was being tested, and then measure its effects in the blood or on behavior once the rat awoke from its anesthesia. If CRH was injected through the tube into the space inside the brain above the hypothalamus, and then a blood sample was collected, a rapid rise in the pituitary and adrenal hormones ACTH and corticosterone soon followed. If a drug that blocked CRH was then injected, it prevented such a rise. And as higher doses of CRH itself were injected, the rats' behavior changed. Gradually, their exploratory movements around a new environment decreased. At the highest doses, the rats stopped exploring altogether and behaved as if ready to escape a threat—when none was present. When drugs that blocked CRH were then injected through the tube, these behaviors ceased, and the rats returned to normal, busily exploring their environment. These studies (done in the early 1980s with George Koob, the team's behavioral psychologist) showed that it was CRH that initiated that classic fight-or-flight response. They showed that CRH was the brain's stress hormone and that the stress behaviors it brought on could be blocked by drugs that blocked CRH. But in a real-life stress situation, which of the brain's chemicals could turn on CRH? The team began testing a variety of neurotransmitters to sort out the workings of the brain's stress response. Then Vale heard about Besedovsky's studies at a meeting and reported them to the group at Salk. Impressed by Besedovsky's studies from the year before, showing that IL-1 stimulated ACTH and corticosterone release into blood, Rivier wanted to know whether the IL-1 was producing its effect by stimulating CRH up in the brain, in the hypothalamus, or whether it bypassed the brain and affected only the pituitary gland.

Around this time, the neuroendocrinologist Robert Sapolsky began to work as a postdoctoral fellow with Rivier at the Salk Institute. Before this, he had trained with his mentor, neuroscientist Bruce McEwen, at the Rockefeller Institute in New York City. There, he had studied the hormonal stress response and the effects of glucocorticoids released during stress on nerve cells in the hippocampus—a part of the brain important to memory function. Sapolsky had found that the high levels of glucocorticoids that occur during stress can kill these memory cells. When he arrived at the Salk Institute, Sapolsky wanted to learn a new technique from Rivier and another of her co-workers, Paul Plotsky. He wanted to learn how to measure hormones coming out of the hypothalamus. Rivier suggested that for his project he measure CRH released from the hypothalamus after IL-1 was injected into blood.

Sapolsky injected IL-1 dissolved in sterile saline solution through a tube inserted into the large neck vein of rats. He then measured blood ACTH and corticosterone levels after injecting different doses of IL-1, as compared to doses of plain saline solution. Just as Besedovsky had found a year before, Sapolsky's team found that IL-1, at even the tiniest doses, caused a rise in stress hormones; the saline did not. The more IL-1 that was given, the greater the stress hormone rise in blood. And when, immediately after the highest dose of IL-1, they injected those antibodies that bound up the CRH and prevented it from reacting, the blood levels of ACTH and corticosterone fell down to the same levels as if no IL-1 had been injected. It was logical to conclude, then, that since the CRH antibody blocked the ACTH and corticosterone rise in blood (induced by IL-1), IL-1 must have made these blood hormones rise by stimulating CRH.

In his studies, Sapolsky further found that IL-1 did not stimulate pituitary cells in tissue culture to produce ACTH. So he concluded that all IL-1's effects on the hormonal stress response came from its effects on the hypothalamus, brain, and CRH. The Salk group published these results in *Science* in 1987. This last finding was important because it was different from what other groups had been finding about the effect of IL-1 on pituitary cells. The third paper in that 1987 *Science* trilogy, for example, had built on another paper published in 1985, which showed that IL-1 *could* stimulate pituitary cells to secrete ACTH. The puzzle was coming together, but some pieces didn't fit perfectly yet.

Jim Blalock, an immunologist at the University of Texas, had been developing a theory that immune cells could produce brain hormones such as ACTH. This was the flip side of the Swiss scientists' theories that brain cells could produce immune hormones such as interleukins. And it was even more controversial. How could immune cells make molecules that were usually secreted by the pituitary? From his studies, Blalock had formulated his own theories of a

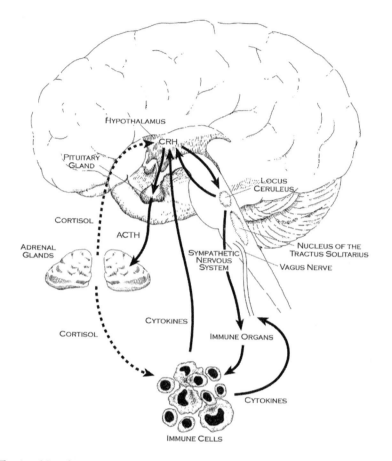

The jumble of arrows in the illustration above show just some of the ways that the brain and immune system communicate. Cytokines, which are made by immune cells, can travel through the bloodstream or signal nerves such as the vagus nerve to activate different parts of the brain. Once activated by the cytokines, the hypothalamus begins to secrete CRH, thus initiating the hormonal cascade of our stress response. The endpoint of this response is the production of cortisol by the adrenal glands. Besides shutting down the brain's stress response, cortisol turns down the ability of immune cells to perpetuate inflammation and produce more cytokines. In this way, the stress response keeps the immune system in check.

two-way communication network between the immune and nervous systems. So his interest was piqued by the news that the Swiss groups had discovered the presence of interleukin-1 in the brain and a rise in

ACTH and corticosterone in blood after exposure to virus or bacterial parts. Blalock chose to test directly whether IL-1 purified from macrophages would stimulate mouse pituitary-tumor cells in culture to make the hormone ACTH. To do this, he added the IL-1 to mouse pituitary-tumor cells. Amazingly, when Blalock added the purified IL-1 to the tumor cells, he could measure ACTH in the culture fluid around the cells in as large a quantity as if the hypothalamic hormone CRH had been added. It seemed that the immune molecule was as good a stimulus to the pituitary cells as was the hormone from the hypothalamus. Blalock published his results in *Science* in 1985—but still, immunologists were not convinced. There could have been many factors mixed in with the purified IL-1 that might have stimulated the pituitary cells. And these were tumor cells, which can do all sorts of strange things in culture. The verdict was still not in proving that normal pituitary cells responded to immune molecules.

About a year later, after IL-1 became more readily available, Captain Edward "Ned" Bernton, an allergist-immunologist at Walter Reed Army Hospital Research Center in Washington, D.C., took Blalock's studies one step further. Instead of using purified IL-1, he added genetically engineered pure recombinant IL-1 to normal rat pituitary cells in culture. He measured all the hormones that such cells usually make, including ACTH, growth hormone, thyroid-stimulating hormone, and others. He did this to find out whether the IL-1 he added affected not just abnormal tumor cells but also nontumor pituitary cells. He, too, found that IL-1 was just as potent as the hypothalamic hormone CRH in making these cells put out their hormones and these were the results that he published in *Science* in 1987.

It took a few more studies by different groups, including the Nobel laureate Julius Axelrod, using different cells and different culture conditions, to prove that IL-1 did stimulate pituitary cells directly. This back-and-forth of findings—sometimes confirming others' discoveries, sometimes not reproducing them—is a course that often occurs in science. Until all the bugs are worked out and all the variables known—such as duration of exposure to a compound, dose, and exact cells and culture conditions—different studies can show different results. This doesn't mean the studies are flawed or not true. It is just the usual zigzag course of scientific research, when one is asking questions on the edge of the unknown and no set recipes are available. In the end, however, it seemed clear from all these studies that

IL-1 could stimulate both the hypothalamus and the pituitary gland: the hypothalamus to make CRH and the pituitary to make ACTH.

Groundbreaking as these studies were, there was still resistance amongst immunologists to the notion that such stimulation of the brain by immune molecules played a role in normal day-to-day situations. Neurobiologists were intrigued but skeptical as well, because it wasn't immediately apparent how such large molecules could travel from areas of inflammation to signal the brain during inflammation or infection. The catch-22 of modern science is that in order to prove something through experiments, you have to manipulate the physiological system—you can't just observe it, you have to interrupt it or replace it. But the very act of manipulating the system changes it artificially, makes it unnatural. So while these contrived experiments proved beyond a shadow of a doubt that immune molecules could signal the brain's stress response, they did not prove that this happened under natural circumstances. Nor did they prove that such stimulation of the brain's stress response played a role in disease. And so, scientists from the disciplines overlapping this field— immunology and neurobiology—remained skeptical, waiting for a naturalistic example that proved the importance of these signaling pathways in disease. And physicians and clinical researchers waited for such findings to be shown in humans.

CHAPTER · 6 ·

When the Brain–Immune Communication Breaks Down

*I*deas are often interwoven and take different forms in different disciplines. Yet what seems obvious in one field may not be accepted in another. In the late 1970s the notion that perhaps glucocorticoids and the brain's hormonal stress response played some physiological role in inflammatory disease had begun to float around in the scientific consciousness. But there weren't enough clear-cut pieces of the puzzle in place at the time to make the theory hold. There was still a strong bias that glucocorticoids were merely anti-inflammatory drugs, not natural immunosuppressants in physiological situations. It took someone with a view of the role of glucocorticoids in the body—an endocrinologist—to lend more credence to the notion that the natural changing rhythms of glucocorticoids during stress might play a physiological role in regulating the immune response and thus immune disease.

That endocrinologist, Alan Munck, proposed this in a hypothetical paper published in 1984. He suggested that the stress response, with its release of steroids, was there to ready the organism for a fight and to protect it from injury. He proposed that the dampening effect of steroids on the stress response formed a logical, built-in brake to the system to keep it from overshooting once the stimulus was gone. Just as glucocorticoids might turn down hormonal stress systems, they might also prevent immune responses from overshooting. He made these predictions based on his work with glucocorticoids and their receptors. If a receptor is in a certain kind of cell, it must be there for a purpose, Munck reasoned. Since immune cells such as lymphocytes contain receptors for glucocorticoids, this might mean that steroid hormones play a role in regulating the immune response, just as they regulate other systems during stress. The receptors would be the molecular lock into which the steroid key fit to shut down immune function when the adrenal glands were turned on during stress.

This hypothesis fit in perfectly with Besedovsky's observations of steroids' suppression of the immune response. Thus, endocrinologists who read Munck's article began to think of immune responses as having some hormonal regulation. But most immunologists were still unaware of this line of thinking and remained skeptical of the possibility that the immune system could be regulated by the hormonal system in day-to-day circumstances. Such seemingly large holes in scientists' collective knowledge have much to do with the blinders we are often forced to wear as a result of focusing so exclusively on one field. It happens in all fields, and, as it turned out, I was guilty of it, too.

I had no idea, for example, when I was experimenting with Lewis and Fischer rats, that halfway around the world in Austria another scientist, Georg Wick, and his postdoctoral student Konrad Schauenstein had seen almost the same phenomenon in chickens—fat chickens that spontaneously developed inflammation of the thyroid gland. I was interested in macrophages, serotonin, and arthritis in mammals, so news of birds and thyroid glands was not likely to come my way. And I was even less likely to go looking for it. But Konrad Schauenstein *had* read Besedovsky's original 1975 article. And although it wasn't until the mid-1980's that he read it, he wondered whether the same studies could be applied to the strain of obese chickens he was studying.

◄o►

In the late 1960s, Randall Cole, a poultry geneticist at Cornell University in Ithaca, New York, noticed that among the thousands of chickens he oversaw, one chicken spontaneously became very fat. He began to try to breed this chicken, to see if he could develop a purebred strain of obese chicken. He bred the fat chicken with a normal chicken, and when the eggs hatched, he found that a few of the chicks were also fat. When the chicks matured, he bred these chickens together, and then their fat offspring together, until all the chicks that hatched were obese.

Cole noticed some other features in these chickens, besides their fatness. He noticed that their feathers were silky and that they had retarded growth. This pattern resembled the pattern of illness in patients with a sluggish thyroid gland. The thyroid, a butterfly-shaped gland in the neck, makes something called thyroid hormone, which controls metabolism, weight, and appetite. People with low thyroid hormone gain weight, undergo changes in the texture of their hair and skin, and develop a deep, hoarse voice. And children with low thyroid function show slowed growth. Perhaps the reason for the weight gain and sluggish metabolism in his obese chickens, Cole reasoned, was a failure of their thyroid glands. Cole asked Ernest Witebsky, an immunologist colleague in nearby Buffalo, New York, to help him diagnose the problem in his obese chickens.

At that time, Witebsky's lab was a hotbed of activity for studies of autoimmune diseases and the thyroid gland. Scientists such as Noel Rose were uncovering mechanisms of thyroiditis—inflammation of the thyroid gland—in primates and humans. And a young Austrian scientist named Georg Wick was working as a postdoctoral student in the lab. Witebsky had assigned Wick to a project studying the organs that controlled chickens' immune function, the thymus and the bursa of Fabricius—the organ, found only in birds, that was named by the founder of Padua's dissecting theater. It had taken more than 400 years for scientists to develop the tools to be able to understand what this organ does, but by the 1970s, immunologists had realized that the bursa was not simply a vestigial organ; it was, in fact, an important source of B lymphocytes—the cells that produce antibodies. (Its equivalent in mammals is the bone marrow.) Wick was learning how to dissect out thymuses and bursae, in order to carry out experiments on antibody production. His training as a physician and pathologist before coming to Witebsky's lab had given him experience in organ dissection that now came in handy in these studies. So, when Cole asked Witebsky to tell him whether the chickens had thyroid disease, Witebsky turned to Wick for his expertise in pathology to answer the question.

Together Wick and Witebsy's team checked thyroid hormone levels in blood samples from these chickens, and found that as the chickens grew and aged, their thyroid hormone decreased, and as it decreased, the chickens gained weight. They also found that the chickens' blood contained thyroid antibodies. Next, the team took samples of the chickens' thyroid glands and examined them under a microscope. The normal thyroid gland is made up of closely packed honeycombs of cells, which form clusters surrounding pools of the thyroid hormone they secrete. In the obese chickens, the clusters were overrun by immune cells—especially lymphocytes. In the obese chickens, the normal architecture of these glands had disappeared, and the neat circles of cells around their hormone pools, like bricks around a garden pond, were replaced with sheets of lymphocytes. What was left of the thyroid glands was scarred and shrunken. Gradually, as they aged, the chickens developed inflammation of the thyroid gland. That explained the fall in thyroid hormone levels and their gain in weight.

When Wick finished his postdoctoral training, he returned to Austria to set up an institute in Vienna to study aging. And he took a few eggs along with him, to start a chicken colony. By then he was the professor, training his own students, one of whom was Konrad Schauenstein. In the early 1980s, after Schauenstein had read Besedovsky's 1975 article showing that withdrawal of corticosterone increased antibody production, he wondered whether something similar would happen in the obese chickens. So he immunized the chickens with sheep red blood cells and measured their antibody and corticosterone responses. The chickens developed antibodies but, surprisingly, no rise in corticosterone. Why? Could it be that they developed thyroiditis because their adrenal glands were functioning poorly? To answer the question, Schauenstein measured corticosterone responses to other stimuli, including to interleukin-1. Again he found their corticosterone response to be low—much lower than levels in their nonobese cousins. Together, his studies showed a very close association between low corticosterone responses and the development of an inflammatory illness of the thyroid gland. Here was an important piece of evidence in proving that blunted adrenal responses in different species can go hand in hand with susceptibility to inflammation.

There was one difficulty with Schauenstein's studies, however. Because the thyroid disease developed spontaneously, and because changes in thyroid hormone change the hormonal stress response, it was hard to know which came first—the altered stress response or the thyroiditis. Perhaps it was some low-grade and undetectable thyroid inflammation that could alter the hypothalamic–pituitary–

adrenal axis. But it could be, too, that the low response of that axis existed before the thyroid inflammation in these chickens, just as it did in our rats before they got arthritis. If that were the case, it would be the low corticosterone response that made the chickens susceptible to thyroid disease. The experiment that would seem to clarify the point—taking out the adrenal glands in the normal chickens—couldn't be done because just taking out the glands wouldn't make these chickens get thyroid disease. Unlike our rats, in which we could induce arthritis at any time by exposing them to strep bacteria, you couldn't take out the chickens' adrenals and expose them to something to induce thyroiditis.

However, those sorts of experiments were soon done in other species that developed other inflammatory diseases. Don Mason, an immunologist studying multiple sclerosis at Oxford University, took out the adrenal glands in rats and then exposed them to a foreign protein extracted from the spinal cord that caused a multiple sclerosislike illness. He then implanted small pellets containing different doses of steroid hormones under the rats' skin. Some rats received no steroid replacement at all. Mason saw the same phenomenon: the rats that had no adrenals and no steroid replacement either died within twelve hours of receiving the spinal cord protein or survived but developed extreme symptoms of multiple sclerosis. Those with full replacement steroids remained well, and those with intermediate doses of steroid replacement developed intermediate symptoms of disease.

◄◦►

At this point, no one involved in this kind of research had actually done detailed testing to determine how their rats had died. We supposed our rats had died of septic shock, because they died so quickly after exposure to infectious products. This was a purely clinical diagnosis, however, based more on observation and assumption than on absolute proof. A different kind of study was needed to prove that point. This was important to do, because it might shed some light on septic shock.

Keith Kelley, an endocrinologist at the University of Illinois, Urbana-Champagne, performed these studies, by exposing septic shock-resistant rats to salmonella bacteria. Salmonella are the sort of bacteria that can be present in spoiling foods and will usually simply make picnickers come down with diarrhea, fever, and intestinal cramps. But a large enough dose in a susceptible host can sometimes produce shock. In his experiment, Kelley cut the stalk of rats' pituitary glands and then exposed these rats—as well as ones whose pituitary stalk was intact—to salmonella. Again, the rats whose pituitaries were cut—whose hormonal stress response had been interrupted at the level of the pituitary gland and therefore couldn't make the pituitary hormone adrenocorticotropin—developed shock. Those with intact pituitaries did not go into shock. They remained resistant to the illness.

Here now was another piece of evidence that the brain's hormonal stress response protects against inflammatory disease, this time for a different illness—septic shock—in response to a different inflammatory trigger—salmonella bacteria. And it was proved by making a cut in yet a different place along the axis, at the level of the pituitary gland. This time the experiments did directly explain what I had seen that night at NIH, when half the rats began quickly dying off. And these studies also explained the episode in the emergency room ten years earlier. The young woman who nearly died of septic shock either had some problem in the glands that controlled her stress response or, more likely, it had something to do with the mysterious drug she'd taken. If any of these factors had interrupted or blunted her hypothalamic–pituitary–adrenal axis, they could have predisposed her to develop septic shock when she developed the urinary tract infection, which exposed her to bits and pieces of bacteria that could cause shock. And these experiments explained exactly why she pulled out of shock when we gave her massive doses of steroids.

—◀o▶—

Putting all of these pieces together—Wick's, Schauenstein's, and my studies, as well as those of Mason and Kelley—the common features stand out more than the different details. The association between blunted hypothalamic–pituitary–adrenal (HPA) axis responses and susceptibility to inflammatory diseases in such different illnesses and different species—in chickens and rats and mice—suggests a common underlying principle of illness. It also suggests that the brain's hormonal stress response is necessary to protect against the massive effects of inflammation, which, without the dampening effects of steroids, can lead to chronic inflammatory illness or even death. If true in so many different animal species, why should these principles not apply in humans, too?

Once again, however, certain parts of the scientific community remained skeptical. Some felt that we had proven that there is perhaps communication between the brain and the immune system—pure recombinant immune molecules could signal the brain—and that this signaling plays an important role in inflammatory disease in animals. But we hadn't yet proven it in humans. Until we did, the skeptics said, it can't be considered important or relevant to disease.

As is often the case, proving it in humans turned out to be much harder than it had been in animals or in a culture dish. The reason is simply that we can't cut nerves and take out glands in humans to see what happens to disease susceptibility. If such drastic measures are ever taken, it's done to cure an illness that is even more damaging. But even then we can't objectively observe the effects of the surgery because the illness itself can change the way the nerve and hormone pathways work. So we can really only observe the system and try to make connections or associations between abnormal nervous system responses and immune disease. The fact that it has taken longer to make definite connections in people has led some scientific critics to continue to claim that the nervous system plays no real role in regulating the immune response and, by extension, that emotions don't really affect disease. This contention once again puts the scientific community at odds with the popular culture, with what all of us have at times experienced of the effects of emotions on disease. Nevertheless, more and more studies are now proving these connections to be important in *human* autoimmune and inflammatory diseases—illnesses such as rheumatoid arthritis, lupus, allergic skin disease, and asthma.

◄○►

Driving north along the Mosel River, through the Alsace region in eastern France, you will pass through rolling hills covered with vineyards and topped with tiny villages and towns with names like Riquewihr and Niedermoschwihr. The names of these old towns, their still-standing walls, and beam and stucco buildings echo their medieval Teutonic–Gallic roots, which made this region always part, and never part, of both Germany and France. If you continue through this bucolic countryside, with its chill mist rising from the river late at night, leaving a dew on grapes in early morning, imperceptibly you cross the border out of France and arrive in Germany at a bustling town called Trier.

Trier straddles the Mosel River and climbs up the hills on either side. It is a modern town that still derives much profit from its wine—a sweet, light, and delicious Riesling. But scattered about the town—around its central thriving square of medieval buildings, eighteenth-century rococo fountain, and modern designer shops and restaurants—are Roman ruins dating from the time that Trier was called Augusta Treverorum, the northern capital of imperial Rome. The route along the Mosel River could well have been the route the Romans took when the emperor Augustus came north to found the town in 16 B.C. By A.D. 306, Treverorum had become the capital of the empire, and the Emperor Constantine held court here in the mammoth palace he constructed. Restored and tinkered with throughout the centuries, it still stands today—now an enormous concrete-and-brick basilica that still gives testament to the imperial power of the city of 2000 years ago.

Also scattered about the town are the ruins of three imperial baths. No Roman emperor could construct a town and palace without also building baths. In the second century, Trier had the empire's second largest, called the Barbara Baths. Today the ruins occupy a square in the midst of the city streets—grass- and hummock-covered ruins that look more like a half-finished, abandoned construction site than the enormous structure they once were. But in Roman times, this was an amazing complex of brick and concrete arches (the Romans invented concrete) 30 meters high, rivaled only by the baths in Rome. The mosaic and marble floors hid honeycombs of ducts and passageways to heat the water and provide the steam for warm rooms, as well as to drain the water for the cold rooms. These cities within cities housed massage rooms, shops, and taverns where the wealthy patricians could spend their days. Although the Roman baths incorporated some of the features of their Greek forerunners— the healing waters, nutrition, and relaxation—they were a far cry from the Greek temples to Asclepius. They were much more centers of social contact, "old boys' clubs," where the powerful could mingle and seal business and political transactions. In their typical fashion of

Although the Roman baths incorporated some of the features of the Asclepions—the healing waters, nutrition, and relaxation—they were much more centers of social contact. The Romans had demystified the temple but kept the physical elements of the cure and added the social dimension of relaxation, perhaps at some level recognizing that social interactions play a role in health. In an uncanny way, they presaged what scientists would later be able to prove, more than 1600 years later, in the very same spot, about the opposite phenomenon: that stress and sometimes social factors can stimulate human hormone responses and affect immune disease.

practical engineering, the Romans had demystified the temple but kept the physical elements of the cure. And they had added the social dimension of relaxation, perhaps at some level recognizing that social interactions play a role in health. In an uncanny way, they presaged what scientists would later be able to prove, more than 1600 years later, in the very same spot, about the opposite phenomenon: that stress and sometimes social factors can stimulate human hormone responses and affect immune disease.

Today, two short city blocks away from the site of the Barbara Baths, is an institute of the University of Trier dedicated to research on psychobiology and psychosomatic medicine, the Forschungszentrums für Psychobiologie und Psychosomatik. Dirk Hellhammer, a German psychologist, built this institute in 1989 to study the effects of stress on immune disease. He had a special interest in studying the kinds of day-to-day psychological stresses to which we are all exposed. And he wanted to be able to define the hormonal responses that transduced the effects of stress on disease. He also wanted to determine whether the principles derived from animal studies were true in humans. Hellhammer soon convinced a husband–wife team of psychologists, Clemens Kirschbaum and Angelika Buske-Kirschbaum, to join him after their return from training in the United States.

Clemens Kirschbaum tackled the problem of developing an easy, noninvasive way to measure stress hormones in humans that did not itself induce stress. Up until then, most researchers measured blood levels of the stress hormones. But even drawing blood is stressful; it causes pain and increases the very hormones that are being measured. So Clemens began to measure cortisol in saliva. First he measured this stress hormone in saliva in all sorts of circumstances—at rest, at different times of the day, and after different kinds of stressful stimuli. And then he compared the levels in saliva to those in blood, to be sure that they matched up. They did. Now the test was ready to be applied to clinical questions.

Meanwhile, Angelika Buske-Kirschbaum asked if there were any immune diseases in humans associated with the same kind of blunted cortisol response as in Lewis rats. To answer the question, she needed to find a group of patients who were well and on no medications, because inflammation and disease and drugs can all change stress hormone levels. She decided to study children with a history of allergic disease, atopic skin rashes or asthma. Buske-Kirschbaum studied them when they were well—in between attacks—and not on drugs. The children, ages ten to fifteen, were asked to bring their best healthy same-sex friend along to the psychology testing lab. Here, in a spare, gray-carpeted, white-walled room, the children were given a task to perform. They had five minutes to prepare a little speech, which they then presented to a panel of white-coated judges, who sat sternly evaluating them at a table opposite the microphone where the children stood to make the oral presentations. Each child was then given a mental arithmetic problem to solve. For most people, these two tasks are potent psychological stressors that consistently cause increased heart rate and cortisol rises. Buske-Kirschbaum collected samples of saliva every ten minutes, by asking the kids to suck on lemon-flavored dental cotton wads. When all the children had been tested and all the saliva samples analyzed for cortisol, she plugged the data into the computer and plotted out the graphs of salivary cortisol over time. They showed that the children with a history of atopic allergic skin disease or asthma had much lower and flatter cortisol responses to stress than did their healthy friends. The Kirschbaums' studies clearly showed that people with a tendency to allergic disease had a lower stress response even when they were well, suggesting that perhaps human proneness to autoimmune disease might come in part from a lack of cortisol response, just like the Lewis strain of rats.

There are many other illnesses associated with a blunted stress response, but two are of particular relevance to our time: fibromyalgia and chronic fatigue syndrome. The symptoms are similar in both: fatigue, mild inflammation, and a flattened stress response. Since, apart from the fatigue, there are very few physical markers for these diseases, patients with these illnesses were at first dismissed by many physicians as hypochondriacs or simply overworked yuppies. There are times when we have all felt our energy flag, when getting going in the morning seems an impossible task. So there is a tendency to expect people with such illnesses to be able to consciously will themselves to snap out of it. But here we aren't just talking about flagging energy—there is an almost immobilizing weakness that makes you fall back into bed the moment you have gotten up. The drained feeling lasts all day, and it comes with aches and pains all over the body. In fibromyalgia the aching is in the muscles, and it comes with tenderness at certain points all over the body—so many spots that at first doctors examining these patients could find no pattern to the pain. This lack of pattern—the vague symptoms that have no apparent physical cause and that don't come neatly matched with any physical signs, as well as the often associated depression—at first made doctors dismiss the illness as not real. But it turns out that these illnesses are real, and in both, the hormonal stress response and other brain chemicals are out of balance. Even skeptics are beginning to accept this now, as over and over again studies show the sluggish stress response. These findings mean something else as well: they point to possible new ways to treat such illnesses—by replenishing the missing hormones, or restoring the balance.

In trying to unravel the causes of both of these syndromes there is once again the problem that in humans, unlike in rats, it's impossible to tell what came first—the flattened stress response or the chronic disease. The reason, of course, is that by the time physicians see these patients, diagnose their illness, and measure their hormones, they are already sick. So it is hard to prove that it was the blunted stress response that made them more susceptible to getting sick, since sickness itself can change the stress response—keep it going too high, or draining it out. Nevertheless, when we put it all together—the findings in animal studies and those in humans—the evidence is overwhelming that the hormonal stress response plays a very important role in either predisposing us to or protecting us from all sorts of inflammatory diseases.

—◄o►—

All these studies make it sound as though the only way the brain controls the immune system is through the hormones of the stress response. This is just one part of the story, however. In fact, it turns out that there are many ways in which the brain talks to the immune system, sending a variety of chemical and nerve signals. Each alone, and all together, are critical in the body's ability to defend against disease. While the hormonal connections between the brain and immune cells were brought to light by immunologists and endocrinologists, these nerve pathways were discovered by another group of scientists working simultaneously, but coming from a different background: neuroanatomy. They also applied their knowledge to the question of the connection between the immune and nervous systems. But trained in the anatomy of nerve pathways, and used to thinking like electricians tracing a break in wiring, these scientists asked a different question: Did nerves feed into immune organs, and if so, did they affect the course of inflammatory disease?

—◄o►—

One of the first to start tracing nerve pathways in immune organs was David Felten. Felten had trained under Wally Nauda, a famed neuroanatomist of the old school, whose pupils learned to map out nerve pathways precisely and in intricate detail using special dyes and microscopes. Felten was interested in the sympathetic nervous system and wanted to map the course of these nerves. But he had the unusual idea of mapping out these nerve pathways not in the brain, as all his peers were doing, but in the immune organs—the spleen, thymus, and lymph nodes. This was such an outrageous idea to other neuroanatomists that he had trouble finding a lab and money and supplies to start his work. Somehow Felten's idea was presented to the MacArthur Foundation, a nonprofit research and arts funding organization that relishes, rather than shuns, high-risk, innovative ideas, and he was given a MacArthur "genius" award. He could now do his research unimpeded by skeptics. He joined forces with his wife Suzanne, also a neuroanatomist.

The Feltens used dyes attached to antibodies to stain the nerves in thin sections of rats' immune organs—spleen and lymph nodes, thymus and bone marrow. These antibodies bound specifically to enzymes that manufactured the neurotransmitters they were seeking. This way, if the neurotransmitter and the machinery to make it were in the nerve, it would stain dark. By staining immune organs for an enzyme that is a signpost for adrenaline-like nerves, the Feltens were able to map the lacey sympathetic nerve pathways interdigitating among beds of lymphocytes in the thymus, spleen, and lymph nodes. So close were these nerves to immune cells that they almost looked, even under an electron microscope, like synaptic nerve-to-nerve connections of the brain. If the nerves and lymphocytes were such close neighbors, the Feltens reasoned, then the nerves must be having some effect on the lymphocytes they touched. But what was that effect?

The Feltens tackled that question by blocking the kinds of nerves they saw, the norepinephrine-producing sympathetic nervous system. They blocked these nerves in two ways, with drugs and by surgical incisions—much the same way as those of us who were studying the HPA axis used drugs or surgery to block the stress hormone axis. And just as we had done, they watched to see what these manipulations did to inflammation.

When they injected the test animals with parts of tuberculosis bacterial walls and oil, a mixture that in rats can cause arthritis, the interruptions of the sympathetic nervous system, whether with drugs or surgical cuts, had different effects, depending on where and how they were made. It seemed as though immune organs needed sympathetic nerves to help lymphocytes mature inside them, to fight off the foreign antigen successfully.

Lymph nodes and other immune organs drain defined regions of the body, just as rivers collect in geographic basins. The drainage basin of some lymph nodes is so precise that in people who have total-body tattoos of different colors you can see a pattern of different colors reflected, as if on a tiny map, on the surface of the lymph node draining each region. This is because the colored dye particles are carried from the skin along the lymphatics to the part of the lymph node that drains that part of skin. Depending on the route of exposure to an infectious organism, or an inflammatory stimulus, a different lymphoid organ—such as lymph node, spleen, or different parts of lymph nodes—will be the first filter point where debris is cleared and where it meets immune cells that can attack. As these cells attack, they grow, divide, and mature to their specialized functions. So, if the nerves that feed into these organs are cut before they enter, they cannot affect immune cells as they fight. Based on these experiments, the Feltens elaborated a theory of the control of inflammation by the sympathetic nervous system. They reasoned that the sympathetic nervous system regulated developing immune responses within immune organs. Based on these findings, the Feltens later showed that the weakening of immune responses that occurs with aging comes in part from a dying off of the nerves that feed the spleen. And this led to their discovery that a drug that makes these nerves sprout and grow again, deprenyl, will reconstitute the weakened immune response in rats whose splenic nerves have died with age or been cut by surgery.

Other neuroanatomists, such as Sue D'Orisio and Karen Bulloch, were also mapping nerve pathways in other immune organs and showing that organs such as the thymus and lymph nodes were rich with many different kinds of nerves. So it seemed that a rich network of many different types of nerve pathways coursed through immune organs and tissues that could become inflamed, such as tissues lining joints. These nerves were different from the sympathetic nerves the Feltens were tracing, and they were certainly different from the endocrine systems we were studying. So here were yet more ways that the nervous system could affect inflammation and the immune response.

Soon David and Suzanne Felten were being asked at meetings how their theories of the sympathetic control of inflammation fit with my observations of the HPA axis. And I was asked how my hormonal theories fit with theirs. The first time I was asked these questions, I wasn't sure how to answer them, except that it seemed to me that the nervous system could be regulating the immune system through all these routes—no one excluded any other. We all agreed. And the more research that has been done, the more it seems this is the case. Indeed, Cobi Heijnen in the Netherlands has now found that children with juvenile rheumatoid arthritis have not only an impaired hormonal stress response but also an imbalance in their sympathetic nervous system responses to stress.

There are many, many ways in which the nervous system regulates immune responses—through hormones and blood-borne routes, through sympathetic nerves in immune organs, and through peripheral nerves in skin and joints and sites of inflammation. The amazing thing about these many routes is that while the body's protective armor of skin and mucous membranes can be breached in many ways and at many sites, no matter how it is pierced or where, there are always nerve pathways or hormones not far off, ready to work closely with immune cells to help them in their fight. Finally knowing all these different routes that connect the brain to the immune system helps us understand just how stress can make you sick and how believing can make you well.

CHAPTER · 7 ·

Can Stress Make You Sick?

————◀◉▶————

We have all at one time or another experienced it: we push ourselves for weeks on end to make a work or school deadline or to care for a sick relative; we overexert ourselves with exhaustive physical training; we go through a lengthy divorce—invariably, when we push our bodies too much in this way, we get sick.

Is there any scientific truth to this? Can stress really make you sick? To answer that question, we need to answer another question first: Do the hormones released during stress change the way our body defends itself? Because unless that ephemeral thing called stress has some concrete way of reaching immune cells, it simply can't make you sick. Yet all those hormonal and nerve pathways that kick in when we are stressed *could* make you prone to sickness, by interfering with the way immune cells cope with disease. Are there differences, then, in the ways individuals experience a stressful event, differences that could lead to variation in hormonal stress responses and ultimately susceptibility to disease?

Professionals in all fields—executives, doctors, lawyers, anyone who must make frequent rapid decisions or perform under pressure—learn to take advantage of their stress response, to use it to bring their performance to a peak. But such individuals also learn to lower their stress response. This may be done subconsciously, or it may be explicitly trained. Doctors in emergency rooms, airplane pilots, stockbrokers, business executives, secretaries, homemakers—anyone who has successfully learned to juggle many tasks simultaneously—learn to assess a situation quickly, break it down to its most manageable parts, prioritize each component, and deal with these in order of urgency. By going through this exercise, whether we have learned by trial and error or have been trained, we are following a pattern of behavior that minimizes our hormonal stress responses. We feel, and then become, more in control.

Imagine for a moment the following scenario. You are sitting on the porch reading; out of the corner of your eye, you see your child run down the driveway and fall while chasing after a ball. You feel a rush of anxiety, but you also feel a rush of power—energy that you didn't know you had. Your heart beats faster, you feel flushed, you sweat. Any fatigue or drowsiness you may have been experiencing before suddenly dissipates. You jump up and run to the child, and, without realizing it, you assess the situation: no cars coming and the child did not reach the street; child moving, crying; bump on forehead but no blood; child reaches for you and climbs into your arms—child is fine, unhurt. As suddenly as your energy surged and your heart sped up, you now feel a wave of relief, as you subconsciously check each item on the list. In a matter of minutes, it is resolved: your stress response has mobilized you, and by systematically assessing the situation, you have then controlled your stress response. But had there been blood, had the child's scalp been cut, without experience and training and the means to treat the cut, you would have continued to have a rapid pulse and to feel increasingly anxious, sweaty, and even faint. Your stress response might have mounted to a point where you were helpless, and, in fact, feeling helpless in a situation makes your stress response spin out of control.

The dose effect of stress—some is good, too much is bad—comes from the biology that underlies the feelings. As soon as the stressful event occurs, it triggers the release of the cascade of hypothalamic, pituitary, and adrenal hormones—the brain's stress response. It also triggers the adrenal glands to release epinephrine, or adrenaline, and the sympathetic nerves to squirt out the adrenaline-like chemical norepinephrine all over the body: nerves that wire the heart, and gut, and skin. So, the heart is driven to beat faster, the fine hairs of your skin stand up, you sweat, you may feel nausea or the urge to defecate. But your attention is focused, your vision becomes crystal clear, a surge of power helps you run—these same chemicals released from nerves make blood flow to your muscles, preparing you to sprint.

All this occurs quickly. If you were to measure the stress hormones in your blood or saliva, they would already be increased within three minutes of the event. In experimental psychology tests, playing a fast-paced video game will make salivary cortisol increase and norepinephrine spill over into venous blood almost as soon as the virtual battle begins. But if you prolong the stress, by being unable to control it or by making it too potent or long-lived, and these hormones and chemicals still continue to pump out from nerves and glands, then the same molecules that mobilized you for the short haul now debilitate you. This dose effect in physiology is called the inverted U-shaped curve, because, if you were to graph it—dosage of hormone versus performance—it looks like an upside-down U. On the rising arm, as hormonal levels increase, performance improves. But then it peaks and, as you slide over the top to the descending limb of the graph, performance fails. Although the point at which performance peaks or fails depends on the type of task that is involved and the kinds of hormones that are being measured, the trend is always the same. And this behavior of the system—some is good, but too much is bad—is not surprising, since it is a general principle of biology that applies to many things—food and drugs, as well as almost any natural substance in your body.

The next question is: What about short-lived stress? Does it affect your immune response and your health? When does stress turn from good to bad, as far as your immune system is concerned? The answers to these questions lie in part in the differences in response time between the nervous and immune systems. The nervous system and the hormonal stress response react to a stimulus in milliseconds, seconds, or minutes. The immune system takes parts of hours or days. It takes much longer than two minutes for immune cells to mobilize and respond to an invader, so it is unlikely that a single, even powerful, short-lived stress on the order of moments could have much of an effect on immune responses. However, when the stress turns chronic, immune defenses begin to be impaired. As the stressful stimulus hammers on, stress hormones and chemicals continue to pump out. Immune cells floating in this milieu in blood, or passing through the spleen, or growing up in thymic nurseries never have a chance to recover from the unabated rush of cortisol. Since cortisol shuts down immune cells' responses, shifting them to a muted form, less able to react to foreign triggers, in the context of continued stress we are less able to defend and fight when faced with new invaders. And so, if you are exposed to, say, a flu or common cold virus when you are chronically stressed out, your immune system is less able to react and you become more susceptible to that infection.

Now think for a moment about the stressful phases of your life—not day-to-day events, but on a longer scale. There are some weeks or months or even years when we may go through more turbulent times than usual. This sometimes has to do with the stage of life, and sometimes just with chance. You may be the mother of an adolescent, learning the difficult process of letting go as the child grows. At the same time, your aging parents may be ill. You are continuously on call for unexpected responsibilities and difficult decisions. Or, at another phase of life, you are the parent of a young child, your first, and you are juggling career and the pressure to succeed. If at such times you are hit with yet another unexpected stress, say the loss of a loved one, you cannot cope. If these stresses are staccato rather than continuous and unstopping, if between stressful events your life settles down to a quiet baseline, then your system will have a chance to recover and be ready for the next assault. But without a safety net, a chronic load of stress accumulates and eventually takes a toll on aspects of your health. This happens because, unless the body has a chance to recuperate, the effects of stress hormones accumulate and build up. These ideas have been borne out in the work of Bruce McEwen at Rockefeller University. McEwen, who has worked on stress and stress hormones for more than thirty years, has shown experimentally that the cumulative effects of high-dose steroids have a different, more long-lasting and harmful impact than if the body produces single, short bursts of them. McEwen calls this compounding of stress effects the theory of "allostatic load."

What kinds of stresses can deplete the body's will to fight? Chronic illness is one. And of course psychological stresses. But there are others as well. Strenuous, unaccustomed, and prolonged physical stress, equivalent to running to your max on a treadmill, for example, but lasting for days, or chronic physiological stresses, such as lack of sleep and food, will all deplete the stress hormone reserves. At first, such chronic stresses keep the response switched on, working at its maximum as long as the stress persists. But if such extremes persist, the response can fail, reach exhaustion, and finally burn out.

◄○►

As we left the choppers .50 caliber machine guns opened up along with .30's. You've never imagined as much havoc. . . .

For four nights and three days it rained and we were awake 90% of the time. No food for five meals, or water. No ponchos for protection. . . . We walked through jungle so thick a machete didn't hardly help. Our bodies took a worse beating than any man should endure. . . .

Few men were bullet casualties, but we had to walk back [through] 10 miles of waist deep water (sometimes chest deep). No choppers because of foul weather. We suffered better than 45% casualties in my platoon because of 'immersion foot'. . . . Some were so bad their feet were a mass of blood. . . .

Have you ever seen a grown man cry? Probably not, well these men were crying while we were returning. It's hard to explain the pain unless you've felt it yourself, but you learn to love a real man over here. These guys won't quit, Doug. We're Marines and we're the best the U. S. M. C. has ever known.

Richard Sutter, a Marine in Vietnam in 1966 and 1967, wrote this gripping description of a "routine" patrol by his platoon, in a letter he sent to a childhood friend back home. Fear of death, excruciating pain, utter physical exhaustion—in a few poignant lines in this letter home, Sutter takes us through the essence of the stresses of war. What forces could possibly be powerful enough to give these men the strength to go on fighting? He tells us this too: they fight for love of friends and for a belief that they are part of something larger than themselves. Sadly, Richard Sutter later died, shot down in an ambush, but he left for us a simple, remarkable legacy.

War is an experience that combines all possible stresses in the extreme, and it does so for prolonged and unrelenting periods: the threat of unpredictable, life-threatening attacks; physical stress and unrelenting strenuous exercise in the harshest environments of extreme heat or cold; lack of sleep, down to three or four hours a night for days on end; the lack of food, eating one meal or less a day for days on end; and the psychological stress of life-depending need for peak performance. In the face of such multiple massive challenges, it is surprising that many soldiers recover without permanent effects on their stress responses. What is not surprising is that some don't recover, continuing to suffer hormonal, psychological, and physical effects long after they have returned to peace and home. And while we don't yet have an explanation for the syndromes of soldiers returning from war—likely many factors and exposures contribute to their cause—the biological effects of massive chronic stress could almost certainly play a role. It remains to test this possibility in experimental settings, to determine whether there are some predictors to tell us which soldiers will and which will not develop later symptoms after being exposed to the same stressful assaults.

The closest controlled peacetime situation that resembles war in its totality of stressors is military endurance training. Army Rangers, for example, are selected for their top physical and mental form, their peak performance in other military tasks. These young men undergo a grueling training period lasting eight weeks, meant to train them to withstand the stress of war. First at sea level, then in mountain settings, they cross-train by running, climbing, and swimming in extremes of temperature and terrain from sweltering jungles to freezing mountains to desert sun and heat. During this time, they average three to four hours sleep a night. Because they eat little, out of anxiety or lack of time, they may lose more than 10 percent of their original body weight.

So concerned was the army about the potential deleterious health effects of such exertions, that they carried out two separate studies to determine whether such massive stress would affect immune response and perhaps susceptibility to disease. Two separate teams of army doctors at the U.S. Army Research Institute of Environmental Medicine in Natick, Massachusetts, and other training centers measured hormone and immune system responses in soldiers before, during, and after this grueling training. There was reason to believe from studies in animals that such prolonged physiological stress might be associated with greater susceptibility to infection. Rats that are not allowed to sleep for prolonged periods, for example, die from overwhelming infections. As we've seen, chronic stress of any sort results in increased cortisol, which should attenuate immune responses and make the host more susceptible to infectious disease.

The teams measured both salivary and blood cortisol and immune system reactivity to common environmental proteins to which we have all been exposed. For this they used a simple standard skin test much like the Tine test that children often receive from their pediatrician. They also measured the ability of immune cells to produce the interleukin tumor necrosis factor, TNF, a molecule that usually enhances immune function as well as other measures of immune system function. They found that by the fourth to sixth week of their training, well into the most stressful phase, the Rangers' cortisol levels had increased and their immune responses had decreased, in some cases to such a low level that the soldiers showed no skin response at all. But three days after the training period ended, and after a period of rest, the soldiers' skin tests and interleukin responses had returned and their cortisol levels had fallen back to the same levels as before the training started. In this setting, then, the increased cortisol and stress hormone responses, and the associated decrease in immune function, was a transient thing, lasting only as long as the chronic stress and returning to normal with a bit of rest. Although these tests still do not tell us whether some individuals might not recover from such stress, they do hint at the direction in which future studies should be aimed. And most importantly, the results of the first study led to a re-design of the training program to minimize the compounding effects of the combined stresses of each training condition. Simply by reversing the order of training, from military base, desert, mountain, and jungle, to base, mountain, jungle, and desert, as well as increasing nutritional intake, the army doctors were able to curtail many of the negative effects of stress on the immune system.

We can ask the same question of people in the general population by exposing them to one component of this stress—the strenuous exercise—and measuring hormonal and immune responses. If you exercise in this way, rather than in a gradated way for training and endurance, you will activate all sorts of neuroendocrine and nervous system pathways. Within ten minutes of beginning the exercise, your hypothalamus starts pouring out CRH; your pituitary gland, ACTH; and your adrenals, cortisol. The cross-talk between hypothalamus and brainstem also turns on your sympathetic nervous system, and the sympathetic nerves, innervating your heart and muscles, pour out norepinephrine, while the adrenal glands secrete adrenaline. Thus your heart rate goes up, oxygen metabolism increases, and eventually fat and glucose stores are broken down.

If you were to measure immune cell function at these times, by counting different types of cells in venous blood or by measuring interleukin production from these cells, you would see that many shifts take place. Certain types of lymphocytes decrease in number and others increase: the helper cells that help to increase antibody production decrease, while the natural killer cells, cells that kill off tumor cells, increase. Cytokine production shifts from a pattern that increases inflammation to one that decreases it. Taking glucocorticoids—steroid hormone drugs such as cortisol—mimics some but not all of these changes. (These are not the same steroid hormones we hear of in the news that make athletes break all records. Those are male sex hormones—chemicals with the same central ringlike chemical structure, but very different atoms hanging off the rings. In fact, while taking male sex steroids will make a person's muscles grow and strengthen, and give him—or her—other masculine features, taking corticosteroids for prolonged periods will weaken strength.) Thus, it seems that all the hormonal systems that kick in during exercise play a role in changing the immune responses during physical stress.

But physical stress is only one component of the stresses associated with immune suppression in something as grueling as Ranger training. And such stress in laboratory settings is short-lived, lasting only twenty minutes or so. What about other forms of stress, such as chronic psychological stress? Is there evidence in average people that chronic psychological stress can change immune function and predispose to susceptibility to disease?

—◄○►—

In the early 1980s, Ron Glaser and Jan Kiecolt-Glaser, a virologist–psychologist husband and wife team at Ohio State University, pooled their expertise together to ask whether the psychological stress of studying for exams had any effect on immune defenses in medical students. In order to answer this question, the Glasers first needed a way of standardizing and assessing the stressful stimulus. They also needed a way to measure precisely some function of the immune system that might be affected by stress. They decided to take advantage of Jan's expertise in measuring stress and Ron's in measuring antibody responses to viruses. They also took advantage of the fact that medical students must be immunized with hepatitis-B vaccine and must receive two booster shots over the course of twelve months. Then they asked whether those students receiving their vaccines during stressful periods of studying for exams showed a lower "take" rate to the vaccination than those exposed when not under stress. The Glasers collected after-vaccination blood from students in whom they had carefully measured psychological levels of stress and measured antibody levels to hepatitis-B vaccine. They found that, compared to nonstressed students, students vaccinated during stressful periods did indeed show lower antibody levels and fewer achieved the clinically significant take-rate of a fourfold increase of antibodies in the blood.

The types of stress that affected the immune function in the Army Rangers are forms of stress that can be considered physical or physiological—exposure to extremes of heat and cold, lack of sleep, poor nutrition. Some of these forms of stress may also have been at work in the exam-stressed medical students—lack of sleep, for example, or poor nutrition. But in both the medical students and the soldiers, there was also an important element of emotional or psychological stress. In both groups, the anxiety of needing to perform at peak in the face of exhaustion and the fear of failure produced stress. For the soldiers, these fears grow from deadly risk in life-and-death situations; for the medical students, such risks, although not life-threatening, were still perceived as potent stressors—fail the exam and you do not become a doctor.

A situation need not entail risk to life to be a real and potent stressor. And conversely, a situation that involves a risk to life is not necessarily perceived by all as a major stress. George Solomon, a psychiatrist at UCLA and one of the early pioneers who tackled the field of psychoneuroimmunology while most scientists were still skeptical, showed this fact in studies performed together with Margaret Kemeny and John Fahey, a psychologist and immunologist at UCLA. Within hours of the 1994 Northridge earthquake near Los Angeles, they measured immune and hormone responses in people who had been at the earthquake's epicenter. They found that while some individuals seemed to respond with high stress hormones and low immune responses, others did not. There could be many factors that contributed to these differences, among them the degree to which each person perceived themselves to be in danger. But all these examples are relatively short-lived stresses. Do different sorts of longer-lasting stresses more uniformly affect immune and hormone responses?

Inescapable exposure to many different stressors simultaneously—a move, full-time work, care of the children and/or household and/or mate—can lead, after many months, to a kind of extreme exhaustion. We call this burnout, and members of certain professions are more prone to burnout than others—nurses and teachers, for example, are among those at highest risk. These professionals are faced daily with caregiving situations in their work lives, often with inadequate pay, inadequate help in their jobs, and with too many patients or students in their charge. Some studies are beginning to show that burnt-out patients may have not only psychological burnout, but also physiological burnout: a *flattened* cortisol response and inability to respond to any stress with even a slight burst of cortisol.

In other words, chronic unrelenting stress can change the stress response itself. And it can change other hormone systems in the body as well. One of the most important of these is the reproductive system. Chronic high stress can shut down reproductive hormones in both men and women: in soldiers undergoing extremes of boot-camp training, people with a heavy caregiving burden, athletes, and those who engage in repeated bouts of extreme dieting. All these situations can stop a woman's menstrual cycle, for example, because stress hormones shut off the monthly surge of sex hormones, which otherwise run the cycle like a clock. In men, such stress also decreases sex hormones, such as testosterone, and so can decrease sperm count and fertility.

With sufficient rest, persons suffering from burnout can recover their ability to make all these hormones, and normal cycles are re-established. We don't know yet whether early menopause can be triggered in a woman close to menopause who experiences such degrees of chronic stress, although irreversible physical changes certainly can result from stress. Women who experience prolonged bouts of depression, in which the stress response is stuck in the on position, do experience permanent changes in their bones—weakening of bones and osteoporosis of the same degree as a menopausal woman twice her age.

In contrast to such chronic, unrelenting caregiver stress—McEwen's allostatic load—there is another form of work stress: the demand for rapid-fire decision making. There is a much more staccato quality to the rhythm of job stress experienced in such professions—frequent, short, but high-intensity bursts of stress. At one time it was thought that professionals in such jobs—without a break and under constant, often uncontrollable, demands—frequently experienced exhaustion, loss of morale, depression, and increased frequency of illnesses.

Imagine a job in which you must be constantly vigilant—even one second of lack of attention might lead to the death of hundreds of people whose lives depend on your moment-to-moment judgments. Imagine that in this job you are working at a small workstation, surrounded by dozens of other co-workers, all of whom are also trying to concentrate on their mission. There is constant movement around you and distracting noise that you must blot out or lose your concentration. You must have lightening quick eye–hand coordination and an ability to react and give commands and directions in response to any shift in the tiny blips you are watching on the screen in front of you. And you must do this job perfectly for hours at a stretch, sometimes late into the night or at the break of dawn. Hundreds of lives depend on your every move. You are an air traffic controller—a member of an extremely high-pressure, high-stress profession. You are someone who is at risk for high blood pressure, stroke, heart disease, accidents, and depression.

In the late 1960s, Bob Rose, a psychiatrist and then an Army Captain at Walter Reed Army Hospital in Washington, D.C., was asked by the Federal Aviation Agency and the air traffic controllers' union to resolve a controversial question using unbiased scientific methods. The question was whether air traffic controllers were at greater risk for developing stress-related illnesses because of their work environment. This was a contentious issue—the union claimed that the stressful work conditions predisposed these workers to illness, and the FAA claimed that this was not the case.

Rose spent one year traveling the country, observing controllers at their workstations, interviewing them and their managers, and assessing their reports of their home environments before beginning the actual collection of data. In all, he studied 400 air traffic controllers over three years to find out what elements of their jobs and their own physiological responses added up to "stress." And he wanted to know what sorts of illnesses they would develop.

Rose's most dramatic finding was related to blood pressure. More than 50 percent of the air traffic controllers studied over the three-year period developed or had high blood pressure. And these men were only in their late thirties. Not all the men developed high blood pressure, however. It was a group of men whose blood pressures were normal off the job, and increased only while at work, who had problems. These men eventually went on to develop high blood pressure even when off the job. In this case, it was the change in heart rate, controlled by the rapid-response sympathetic nervous system, that, more than cortisol, matched immediate changes in work stress.

Many of the controllers also experienced depression and anxiety disorders. The single factor that was linked most closely with such illnesses was something that could not be measured in urine or blood or with a blood pressure cuff. It was the pervasive sense among all these workers of alienation and abandonment by their employers that mapped most closely to these illnesses. Those who perceived their work environment to be uncaring and unsupportive were at greatest risk. And in this group, it was the men who felt the greatest sense of distance from their employers, the greatest sense of "we against them," who developed illness.

Rose's initial study of air traffic controllers was completed more than ten years before the air traffic controllers' strike and the mass layoff of thousands of controllers in 1983. At the time of the strike, a large percentage of controllers found themselves suddenly out of work. Here were men in their prime, heads of households, highly trained, and skilled in a very specialized profession. Suddenly, and virtually without warning, they lost their jobs, with no recourse or possibility of returning to their profession. Many experienced clinical depressions during the first year after their layoffs. Some fell back into old habits of drinking to mask their problems. Eventually, most found new and productive occupations, and put the strike and depressions behind them. But some did not.

Ten years later, in 1993, Rose tracked down and interviewed as many of the controllers as he could find—in all, about two-thirds of the original group. He found, unlike twenty years before when the men were in their thirties, that a higher number of these men in their fifties had illnesses such as cancer, heart disease, alcoholism, and depression. This was, of course, to be expected: These illnesses occur more frequently in men in their fifties than those in their thirties. But when Rose went back and looked to see which measures, if any, predicted the development of these illnesses, he found something unexpected. More important than any of the changes he had been able to measure twenty years earlier in blood, urine, or blood pressure, Rose found that it was the psychological factors that best predicted development of these illnesses twenty years later.

━◀○▶━

Let's return for a moment to the thing that began all of this: What is stress? One part of it is obviously your body's response, the many hormones and chemicals that we've just discussed. This is the part of stress that directly impacts the immune cells' ability to fight. This end of the process, the hardwiring and chemicals that make it go, is like the motor, driveshaft, wheels, and axles that drive a car. But what about the ignition? What turns it on? As important as the motor is what happens between the event itself and the perception of the event.

For every individual exposed to an event, there is a different interpretation of its stressfulness. Recall the situation of seeing the child who runs down the driveway, falls down, and bleeds. A doctor or a nurse or an emergency technician will feel less stressed than someone unfamiliar with such injuries. In the emergency room setting, the health professional's stress threshold has been raised by knowledge and experience and ability to act. A lot of blood in this case isn't life-threatening, a few stitches usually suffice. But if the doctor or the nurse is presented with the same situation, but without access to suture equipment—say, on a busy highway—it takes less for their stress response to kick than it would in emergency room or clinic, where they can staunch the bleeding right away. It takes a lesser stress to turn on the body's stress response when we don't feel in control. Our perception of stress, and therefore our response to it, is an ever-changing thing that depends a great deal on the circumstances and settings in which we find ourselves. It depends on previous experience and knowledge, as well as on the actual event that has occurred. And it depends on memory, too.

A memory is not a threat—it cannot kill or harm, and yet a memory of a stressful event can turn on the stress response almost as much as the original event itself. This is similar to what we experience when watching a movie or playing a video game—we know these are not real and yet we feel anxiety or fear. In the case of memory, this happens because there are many nerve pathways leading from the brain's memory centers to the hypothalamus that can trigger the stress response. One of these memory centers is called the hippocampus, that part of the brain named for its sea horse–like shape. Another is in the frontal lobe, at the front part of the brain.

One difference between real and virtual threats is that the more often the memory or the movie or the video game are re-experienced, without harm actually occurring, the weaker the body's physiological responses soon become. Because we quickly learn that these events will do no harm, the stress response to such virtual experiences eventually extinguishes. Hence the desire by some, where actual video games are concerned, to go on to try new and more stimulating videos, to see new movies with more violent special effects. But hence, too, the thankful relief as time distances us from a real-life traumatic experience. Time heals all wounds because we forget.

There are some unfortunate individuals however, for whom the memory of a massive trauma never fades. These persons repeatedly re-experience the memory of the event in all its power, with all the physiological, nerve and hormone responses that it initially evoked. Even the tiniest reminder of an event, a glimpse of a person or an object that bears even the slightest resemblance to some part of the original, can trigger the whole cascade in its most fulminating form. Soldiers in every war have experienced some form of this syndrome, given different names in different eras: Da Costa's syndrome in the Civil War, shell shock in World War I, battle fatigue or "disordered action of the heart" in World War II. Today this syndrome is called post-traumatic stress disorder, or PTSD. It is seen in Holocaust survivors as well, and in civilians exposed not to war but to an equally horrific trauma—bombing, fire, or rape victims, for example. If a woman was raped at dusk beside a tall boxwood hedge, seeing a hedge of approximately the same height at dusk, or smelling boxwood, can later trigger her whole stress response. Being forced to recount the events, to visualize them in memory, can trigger it as well.

Rachel Yehuda, a psychologist at Mount Sinai School of Medicine and the Veterans Administration Hospital in the Bronx, realized when caring for Holocaust survivors that the adult children of these persons also exhibited abnormal responses to stress, even though they themselves had never experienced Holocaust or war. When she began to test the hormonal responses to stress of both survivors and their first-degree relatives—children or siblings—she found that they all had higher than expected cortisol rises in response to stress and lower resting levels of stress hormone rhythms throughout the day. This could be something learned. The survivors, when the trauma was still very fresh, may somehow have subconsciously taught their children to respond in a certain way to stressful stimuli from a very young age. But just as likely is the possibility that these patterns of stress-responsiveness are inherited. In that case, the people who go on to develop PTSD if exposed to a terrible stress would be those who have inherited a stress response that doesn't extinguish. Whether learned or in the genes, or a little bit of both, for victims of this illness distance from the event does not diminish the body's physiological response. They are those for whom memory stays alive.

Stress need not be on the order of war, rape, or the Holocaust to trigger at least some elements of PTSD. Common stresses that we all experience can trigger the emotional memory of a stressful circumstance—and all its accompanying physiological responses. Prolonged stress—such as divorce, a hostile workplace, the end of a relationship, or the death of a loved one—can all trigger elements of PTSD.

Now imagine the following scene: you wake up refreshed and happy, then relax and read the newspaper over coffee, a sweet peach, and a roll. You feel happy and secure as sunlight streams into your kitchen. Then you leave for work. Work is a hostile environment where day after day your boss disparages you inappropriately; where there is the threat of job loss because of downsizing; where there is inadequate infrastructure to support your productivity; where the physical surroundings are cramped and noisy; where you are not valued for your full worth. As you drive toward the office, your mood gradually deteriorates. As the distance shortens, you become more and more tense. The moment your car passes into the parking lot, you feel a rush of anxiety, increased heart rate, a mild flush. To add insult to injury, there are no parking spots because company policy reserves spots only for those of higher rank. You park anyway in the full knowledge that when you return to your car at the end of the day there will be a parking ticket on the windshield. You step out of your car and walk toward the elevator, anxious, angry, demoralized, and dreading the start of the workday.

Or maybe your workplace is heaven: a bright and airy office, supportive co-workers and boss, exciting projects, enthusiastic management that values its workers—but back home, your life is imploding. You are in the midst of a nasty divorce from a controlling spouse, one who had emotionally or physically abused you during the marriage. Day after day for months on end, the attorney of your soon-to-be ex-spouse, known as a pitbull divorce lawyer, a basher who takes pride in destroying lives rather than salvaging what is left of the family's spirit, uses grinding tactics to wear you down. His repeated prying questions are designed to trap you, set you up against yourself. He waits a few days, then escalates the legal demands, threatening subpoenas and depositions. His threats come in waves, so that as soon as you have regained some balance, he hits you again. You feel like one of those inflatable plastic punching toys that is slapped down the moment it pops up again. And the threat this attorney is using to break your spirit is loss of custody of your children. While the target of such attacks, you may experience palpitations, flushing, an urgency to defecate every time the phone rings or whenever a letter is delivered to the door. You may have repeated nightmares—losing your children, searching for them but not finding them. You may wake up in a cold sweat. You may even continue to experience such physical symptoms and anxiety long after the divorce is over and a settlement has been reached.

You are experiencing a shadow form of the elements of PTSD. The trigger to these symptoms need not be complex, if the initial event was severe enough. Sometimes a single visual element can expose a shard of memory that evokes a physiological response. For example, something as innocent as a lawn marker for a house address—a gray stone with the address painted on it—may, after the death of a loved one, remind us of a gravestone, and for a transient few seconds bring on the rush of hormones and despondent feelings that we experienced when the loved one died.

The standard list of life situations that can act as powerful stressors, quoted in textbooks of psychology and psychiatry, includes loss of a loved one, divorce, loss of a job, and moving. What an odd and rag-tag list of situations to juxtapose—something as mundane as moving next to something as profound as death. Yet moving *is* a major stressor. What could be the common elements that link all these situations? One is certainly loss—the loss of someone or something familiar. Another is novelty—finding oneself in a new and unfamiliar place because of the loss. Together these amount to change: moving away from something one knows and toward something one doesn't.

The stress of loss comes in part from a kind of grieving. Holes are left in your memory in the places where the familiar used to be. You have moved to a new house. You reach for a book on the shelf and it's not there—the shelf may not be there either—the room is different. You know exactly where you put the book, you see it in your mind's eye, but the reality has changed. These losses are the same but less intense, less immediate, less emotional than in true grieving.

Certainly the loss of a loved one—especially a spouse, a child, or a parent—causes the deepest levels of grief most of us will ever know. The presence of the deceased seems to linger. You turn to the spot on the sofa where your wife used to sit, beside you. You try to feel her warmth, but instead there is a coldness there. You try to reconstruct her face, but can't; it is years now since she died. The edges of the memory have faded, gradually erased, and there is nothing you can do to bring it back—the physical is no longer in the world to reinforce the image in your mind. This jarring, painful reality breaks through what was otherwise a background music of awareness that something was missing from the order of your world.

It takes time to reconstruct a new memory, a new place in your mind that matches the new shape of your physical world. And it takes time for the old memory to fade away—to fade until the intensity of the memory is less than the intensity of the reality that evoked it. And until this happens, each time you are reminded of your loss, a whole set of emotions is triggered—from anxiety at the new world order to sadness at the loss. Grieving is all these things: unlearning of old memories, relearning new ones.

In a way, our whole world is represented in our memory—peopled in familiar space, like stars hanging in a firmament for which we have drawn the bounds and filled the background. We navigate this territory of our mind more often in the day than we do our real surrounding world. We go there during those cracks and crevices of time when our thoughts wander from the moment. Driving down the street, stopped at a light, at our desk at work, at home in the kitchen preparing for a meal, in bed, drifting off to sleep: our thoughts shift to this other world that we carry with us all the time. And in some ways it is this world that matters most—the relation of each person to each other person, the relation of each person to their rightful place. When any of this changes, stars extinguish and patches of background tear apart; the shifting scene is unfamiliar, even frightening. Until we make sense of it, reestablish order in this internal map and relearn connections, we cannot find familiarity and peace.

An unfamiliar environment is a universal stressor to nearly all species, no matter how developed or undeveloped. Even rats, taken out of their home environment and placed in a clean, brightly lit cage, will show signs of anxiety and stress: they will decrease their explorative behaviors, even freeze, and defecate more—all fight-or-flight behaviors controlled by the stress hormone CRH. Anyone who has had a pet small enough to hold in the hand—a hamster, gerbil, or toad—will have noticed this tendency to urinate or defecate when it is first picked up. If you were to measure stress hormones at this time, you would find them to be high. (In fact, this was the test we had performed to show that our high-CRH responsive Fischer rats behaved differently in stressful situations compared to their low-CRH Lewis cousins. The low-CRH Lewis rats explore their new environment, while the high-CRH Fischers tend to stay in one corner chasing their tails.) But let the rat acclimatize to its new environment, or repeatedly put it into the same, now less new, cage, and the signs of anxiety, as well as the hormone levels, will decrease over time. It gets used to the new environment. It learns.

When first placed into a new environment, say a maze shaped like a plus sign, the rat runs furiously in all directions, sniffing, zigging, zagging, looking up and looking down. With this activity, the rat is memorizing visual cues to construct an inner map of its environment, just like that inner map of our sensory and motor selves that Wilder Penfield mapped out when he placed electrodes into the human brain. If you were to place electrodes in the rat's hippocampus and record activity in those nerve cells when the rat is placed into the new environment, you would see an amazing sight recorded on the computer image: the active brain cells are laid out in the exact shape of the space that the rat has explored—they are laid out in the shape of the maze's plus sign. And the cells that are most active are the ones that record the place the rat is exploring at that moment in time.

This anxiety about novel environments that keeps us vigilant is controlled by two parts of the brain—a part of the brain that controls memory, and another part that controls anxiety. The part of the brain that integrates memory of our spatial world is the hippocampus. The part that controls anxiety is the fear center deep within the brain, the amygdala. Both of these parts of the brain have connections to the brain's stress center.

If the hippocampal memory part of the brain is damaged, learning will not take place. And anxiety will not decrease after repeated exposures to the new environment. We don't yet know all the wiring that makes this work, but somehow a cue in the environment—a certain tree in the woods, a building, a stoplight, a painting in a room—gets temporarily connected to the amygdala. It could be that we actually reconstruct a spatial image within our hippocampus, or it could be that the hippocampus works more like pixels in a computer—acts as a transient connecting station that links two otherwise unconnected spots. For a time, the unfamiliar images get linked to the amygdala, until, with learning and familiarity, the links are lost.

From an evolutionary point of view, this behavioral response is clearly adaptive. An animal in a new environment must map out the new territory, quickly determine where its predators may be lurking, define escape routes in case of attack, and commit them to memory. All this requires vigilance, focused attention, a readiness to flee—behaviors that are programmed and induced by the brain's stress hormones. So in this setting, stress is not only "good"; it is necessary. Of course these feelings of stress in a new environment are uncomfortable. If they weren't uncomfortable the animal might not be motivated to do something about the situation to change it and protect itself. This is adaptive, too. For different reasons then, both loss and novelty—and certainly both together—can activate the stress response.

This same sequence of events occurs with people. Walk into an unfamiliar room, hike in unfamiliar woods, arrive in a new city. Your eyes are scanning all around—taking in first the general impression and then the details. Unfamiliar surroundings bring on anxiety. Repeated visits to the place, repeatedly seeing the same tree on the hiking trail, the same picture on the same wall in the room, the same stoplight on a street—all help decrease the anxiety of the unknown, help build a memory and a sense of peace. The memory that is built like this is not one of a word or of the object alone. It is a memory of the object in its surroundings. This kind of learning is called spatial learning. And if there is a missing piece in this otherwise familiar space in memory (due to the loss of a loved one, moving, or divorce), the hole is all the more apparent. And because the brain pathways activated by the loss connect to and activate the stress response, the anxiety and loss that accompany change can trigger all the hormone and nerve responses that can then alter immune responses.

Is it possible that there are some among us who, because of a particular load of genes, may experience stress in different ways or may be less able to extinguish painful memories or responses? If we go back to the rats—those strains bred first for their resistance or susceptibility to inflammatory disease—we know they do have different stress responses. One strain, the Fischer rats, pours out stress hormones at the slightest perturbation, while the Lewis rats, seem to wander through life apparently imperturbable, stress hormones flat no matter what occurs. In fact, rather than becoming anxious, in stressful circumstances these Lewis rats will often go to sleep. This is not a terribly adaptive behavior for a rat in the wild, since if attacked by a predator, a rat needs to be able to mobilize all its defenses quickly and run, not curl up and go to sleep. This strain of rats probably survived only because they were bred in the protected surroundings of the research lab.

Our response to a sudden unexpected event—that is, our level of surprise—is a very primitive reflex that is crucial for survival. It is an important element of the stress response, even if the event itself is not severe. A startling low-grade noise can act as a mild form of stress for rats, or mice, or humans. The startle response, as it is called, is a reflex all mammals have, as simple in numbers of nerve connections as the knee-jerk reflex. It is a protective reflex that readies you for fight or flight. If someone comes up behind you and claps their hands or you hear a sudden loud pop or backfire of a truck, almost instantly you will stiffen and blink. Fear can make your startle more intense. If you thought the truck's backfire was a bullet, you will do more than blink—you will jump. A rat will do the same: learning that something should be feared by previous or repeated association—by conditioning—can intensify the startle response.

In humans, the strength of the startle response can be measured by gluing a small electrode to the eyelid. How hard you blink correlates with the degree of fear you feel. In a rat, the strength of a startled jump can be measured on an instrument much like a digital weighing scale. The reflex nerve pathway that governs this response starts in the ear and is triggered by a sound. Electrical signals lead from the ear to the centers in the brain that interpret sound—the auditory cortex. But some signals bypass the cortex and travel through three short nerve connections deep within primitive parts of the brain stem. From there they move on to nerve paths in the spinal cord that lead to muscles of the eyelids, trunk, and limbs. Like any reflex pathway, the startle reflex is automatic—something we are born with. But learning can modify this pathway—intensifying it, for example. If by previous experience we are startled by a fearful event, fibers from the fear center in the brain, the amygdala, can connect to these nerve routes. From there, the stress hormone CRH (which comes from the brain's hypothalamic stress center and triggers the pituitary gland to squirt out ACTH and the adrenals to make cortisol) can increase the amygdala's responses. So in a learned, fearful situation this stress hormone changes—amplifies—a reflex response. The startling noise can activate the stress response and the hormonal stress response can make the startle increase.

When we first discovered the difference in stress responses in Lewis and Fischer rats, we also tested whether they would startle differently when exposed to a sudden noise. They did. But, surprisingly, it was the low-stress Lewis rats that startled more than the Fischer strain. Stafford Lightman, a British neuroendocrinologist, later decided to use sophisticated computerized laboratory equipment to study what happens to stress hormones in these different strains of rats when they startle. Lightman was able to continuously collect tiny volumes of blood, in which he monitored the concentrations of stress hormones throughout the day. When the rats were resting, feeding, or just wandering about their cages, they showed the expected regular "pulses" of the hormone corticosterone.

At different times of day, there is a regular pattern of ebb and flow of all our stress hormones. This is the background music of our stress response that goes on regardless of those external stresses to which we are exposed. What appear as smooth waves of hormones surging and falling when we measure them in blood are actually made up of many minipulses of each stress hormone. (This pulsating pattern of stress hormones, which occurs in all species, comes from the rhythm of the hypothalamus, although we don't completely understand why the hypothalamus "beats" like this. It has something to do with the brain's other biological rhythms, which begin in the clock center of the brain.) At times when these pulses increase in frequency or in size the hormones measurable in blood also increase. In humans this happens in the early morning when we wake up, and plasma cortisol and ACTH reach their peak. As the day wanes, the pulses spread gradually apart and decrease in size again, until, in late afternoon and early evening, they bottom out, to stay low until they surge again in early morning, sometime around dawn. The nadir of these hormones coincides with the late afternoon dip in energy we often feel that drives many of us to seek a cup of coffee in order to refuel. In fact, one effect of the caffeine in your coffee is to stimulate the release of hypothalamic CRH—in other words, it gives your stress hormones a jolt. If, on the other hand, the early morning peak occurs too soon, as happens when we are experiencing chronic stress, or some forms of depression, we will wake up—bolt upright, heart racing, ready for the day—but far too early, before the dawn, at exactly the time the stress hormone surge occurs in the middle of the night.

Rats have the same rhythm to their stress hormones, except that their circadian rhythms are exactly reversed: they bottom out in the morning and peak at night (not a surprising pattern for a nocturnal animal). As expected, Lightman found this same hormone pattern in his rats, each one rising from baseline to a peak over the course of about one minute. Then, by monitoring the hormone levels and exposing the rats to a quick burst of noise either at a low point between peaks or on the way down from a peak, Lightman discovered something unusual. Most rat strains, when exposed to noise during a low, responded with a burst of stress hormone, but, if they received the noise during a peak or on it's downward stroke, they did not respond at all; they were refractory to the stress.

But Lightman discovered something else: in the extremely stress-responsive Fischer rats (these are also the rats that are resistant to arthritis), there was no refractory period. They responded to a noise experienced during a naturally occurring peak with the same amount of stress hormone that they would if exposed during a naturally occurring trough. And with repeated noise, their stress response did not extinguish, as it did in the non–stress-responsive strains. They responded to each noise as if none had come before. From Lightman's experiments, then, it would seem that hyperresponsiveness to stress and the inability to become accustomed to a stress in these purebred rats might be genetically determined and perhaps present from birth.

These two rat strains are genetically selected, one showing hypersensitivity to stress and on the other, sluggish hormonal responses. The hypersensitive responders are protected from arthritis but are easily triggered to respond to stress, while their cousins are the opposite—although protected from the effects of stress, they are not protected from diseases such as arthritis. Are there genetic situations in humans that are comparable? Situations in which different groups of people show a greater or a lesser hormonal response to stress? And if there are, do these people show different tendencies to develop autoimmune or inflammatory diseases? We are beginning to discover groups of people predisposed to inflammatory disease whose response to psychological stress is also blunted. (These are the children with allergic skin disease or asthma whom Angelika Buske-Kirschbaum found to have low levels of saliva cortisol in response to public speaking and mental arithmetic.) We don't yet know if these individuals also experience stressful situations differently from the high-cortisol responders.

So there is a yin and yang to stress hormone responses. On the one hand, the degree of stress responsiveness you begin with can predispose you to or protect you from disease. On the other, just being stressed, and being stressed for a long period of time, can change your susceptibility to disease. But instead of a susceptibility to inflammatory or allergic diseases, such prolonged stresses predispose to the opposite—infectious diseases. In the first, the immune response is unchecked, and in the second, it is held in check too much.

◄○►

So, stress *can* make you sick because the hormones and nerve pathways activated by stress change the way the immune system responds, making it less able to fight invaders. There are many parts to stress between the thing that happens to us and those hormones' effects on the function of immune cells. There is our perception of the event, and there is our genetic set-point of our stress responses. Some of us are high and some of us are low stress responders. Certainly we can do something to change our perception of an event as stressful, tone down our physiological responses to that stress, and minimize stress' effects on disease. Here memory and learning play a role—memory of what was or memory of what should have been. But even learning and familiarity, can't entirely override the degree of stress-responsiveness we are born with.

In all these stressful situations there is another element that contributes to the perceived stress. In each of these settings, besides physical, physiological, and emotional stresses, there are interpersonal relationships, which in some cases contribute to, and in other cases can buffer us from stress. In many ways, relationships can be the most powerful stressors that most people will encounter in their waking, working lives. And they can be the greatest soothers, too.

CHAPTER · 8 ·

Connecting to Others

Relationships and the Course of Disease

————◀○▶————

Somewhere in our brains we carry a map of our relationships. It is our mother's lap, our best friend's holding hand, our lover's embrace—all these we carry within ourselves when we are alone. Just knowing that these are there to hold us if we fall gives us a sense of peace. "Cradled," "rooted," "connected" are words we use to describe the feeling that comes of this knowledge; social psychologists call this sense embeddedness. The opposite is perhaps a more familiar term—we call it loneliness.

Thus a person, sitting by herself in a room, may appear to others to be quite alone; but that person, if embedded, will have a world of relationships mapped inside her mind—a map that will lead to those who can be called on for nurture and support in time of need. But others, the Gatsbys among us, might be among a crowd of dozens and yet feel very much alone. Many pieces of great literature have in fact tapped into this sense of disconnectedness. Our sense that powerful forces beyond our bodies link us to others is so ingrained that we use phrases such as "ties that bind," "family ties," and "bonding," to describe those intangible connections. And the emotions they evoke are among the greatest forces that affect our hormonal, our nerve chemical, and our immune responses—and through these, our health and our resistance to disease.

The roots of this sense of attachment to others in our world starts very young. A small child, a baby chimp, a kitten—all at first cling to their mothers. And mothers of all mammalian species, as long as the infant is still helpless, remain close and present to nurture. As the young develop, they gain skills that permit them to venture out and away from mother in ever-increasing circumferences. But, as if brought back by a tensely coiled spring, the young return for warmth and succor when they reach the edge of that secure and charted world. What helps these young leave their mother's lap—the certainty of warmth and food—to explore the unknown, possibly dangerous world around them? At first, a very young child will carry a physical reminder of mother's embrace: a security blanket, a favorite toy, something soaked with all the smells of home and love. A freshly laundered object won't do. And then, as the child gains confidence and skill, the physical object gradually gets left behind, dissolves into a memory, and becomes part of the mental map within.

In different iterations we do the same throughout our lives with all our significant relationships. At first it is too painful to leave our loved one's side, but soon we must. And so high school sweethearts in different generations have worn pins or rings, carried lockets or stuffed animals, or kept the dried petals of a corsage to remind them of their love. The engagement ring and wedding band have the power in an ounce of gold to evoke the memory of the beloved. After the death of a spouse or parent, the grieving partner or child wears the ring in a chain around her neck, a physical reminder that the loved one lives on inside the mind. These reminders can evoke a sense of peace as well. But if the relationship was stormy, if the marriage ended in divorce, that same ounce of gold can elicit such revulsion, such a shudder, that it is thrown into the fireplace ashes, pawned, or melted to a different shape in an attempt to blot the memory out.

We are all tethered to our social worlds by invisible but steel-strong wires. These wires are the signals we constantly send to those around us, which they decode and to which they respond: nonverbal cues, body language, tone of voice—the subtext of our verbal communications. In his poem "The Ecstasy," John Donne, the sixteenth-century English metaphysical poet, captured the progression of these social cues from physical to emotional:

> Our hands were firmly cemented
>   With a fast balm, which thence did spring;

Our eye-beams twisted, and did thread
Our eyes upon one double string; . . . .
Our souls (which to advance their state
were gone out) hung 'twixt her and me.

What does it take to build a social bond? We are connected to our social world through all our senses. Each adds a new layer of richness and depth to the emotional reactions that world evokes. We receive signals through each sense, and we use each sense to send such signals to others. It takes many different kinds of sensory signals and our full range of emotions to create that bond. And it takes memory, too.

"Life is . . . made up of exquisite moments," said Oscar Wilde. Indeed, if a movie recounted every real-time minute of the day, it would be impossibly long and excruciatingly boring. A good screenwriter or author or playwright selects the moments that are to be described in detail and strings them together to move the plot along. So, too, does our brain make such selections in stringing together memories. A relationship is built of strings of moments that our mind has pulled out from where they were stored in memory, moments and memories that come with emotions attached. Memories, spliced together like this in a seamless thread, make a relationship seem continuous and whole. So, after not seeing a childhood friend for years, we can pick up where we left off, as if no time at all had intervened. In this way, too, relationships can be sustained in thought during long absences—parents away from adult children, long-distance lovers, commuting husbands and wives. But the same capacity of the brain to forge this chain of memory can lead to difficulties in a relationship if one member evolves past where the other's memory left off. So, a child leaving home for college, who left still on the verge of adulthood and returns an independent adult, will encounter a parent's resistance when the person who steps back into the parent's memory is not the same as the one who left. It takes a period of adjustment on both sides to set the chain evolving back on a new course.

The reason we piece together relationships in this way is a result, in part, of how the regions of our brain that govern memory work. The front part of the brain, the one that does the day-to-day, on-line job of keeping track, is like a bin into which we can reach and pull out bits and pieces of memories that we then string together to make a whole. However, the bits and pieces come from different places in the brain that have received these signals first, those parts that are the direct receivers from each sense: vision, hearing, touch, smell, and taste. Other bits come from emotional centers that have added pleasant or unpleasant charge.

Thus, whether we are with others or alone, social interactions or just the memory of them can trigger these emotional responses. Go back to that map inside our heads, the map of our social landscape that mostly stays in the background of our lives. At times, one small corner of that map can swell and grow, reverberate and suddenly seem to take over our entire world: we fall in love; we are abandoned; we become envious; we hate. The persons who are the object of such feelings can take on gigantic proportions in our minds and dominate our whole social and emotional outlook, coloring every corner of our lives, until, through monumental effort, or simply through gradual erosion of time, they recede again to their rightful place and size. In these settings, one need not actually confront the objects of the emotions—they occupy a place in one's memory or consciousness that itself can elicit the same emotional response as if the person were present.

Visualizing a loved one, remembering that person, looking at a picture of that person—all these can all bring on the soothing emotions of pleasure or the activating emotions of sexual arousal. Visualizing, remembering, or seeing a picture of one who is hated or feared can bring on all those sets of emotional and accompanying physiological responses. Scientists who study the emotions take advantage of this fact when they use pictures to trigger emotional responses. A standard set of pictures was developed more than two decades ago by the psychologist Peter Lang to bring on specific emotions in the laboratory setting. When Lang developed them, he was able to show these pictures to normal people and to measure their increased heart rate or other physiological responses to emotional arousal. Although not proven by systematic studies, it is the impression of some researchers that today pictures of horrible events seem less able to evoke strong responses in test subjects than when the set was first developed. Indeed, Lang himself constantly revises and updates the set to maximize its emotional impact. Perhaps we have become so inured to such images—accustomed as we are to shocking television, films, and news photos—that a simple set of pictures can no longer move us. Or perhaps the unidimensionality of the pictures, the fact that we know they are just pictures, protects us from an emotional response. Today, short of being confronted by the actual object of our emotions, the most universally powerful evoker of emotions can be found in motion pictures. The difference between an average and a great film, an average and a great director or actor, lies in their ability to make us genuinely feel. And in order to accomplish that, the film must be evocative, the characters rich with many dimensions, even evolving over time. Great books can do this, too, sometimes even more powerfully than film, because in a book the author has the advantage of the reader's imagination. The imagery in a book can tap into an already held image that the reader can visualize, an image that when visualized can evoke a powerful internal emotional response.

Why is imagery so powerful? Could there be something about the way the brain reconstructs images that makes a thing seem real, and so connected to emotions? We know that different parts of the brain become active depending on whether we visualize an object or actually see it. Steve Kosslyn, a psychologist at Harvard University, showed people a complex shape and then asked them to visualize it—see it in their mind's eye—then turn it around and look at it from different angles. All the while, the subjects were monitored by a PET scanner, an instrument that detects changes in brain blood flow and metabolism. When the subjects looked at the shape, the part of the brain where blood flow increased was the part that receives signals from the eyes and integrates those signals into the shapes we perceive—the visual cortex in the back part of the brain, called the occipital lobe. But when the subjects visualized the same object without seeing it, it was spots in the parietal and frontal lobes that lit up. The frontal lobe is one part of the brain that integrates memories "on-line"—the part that controls our "working" memory function. The parietal lobe is essential for tasks requiring imagination and the perception of objects in space. It is also the place where dreams come from. If blood flow to the parietal lobe is cut off, and this area of brain tissue dies, as sometimes happens during a stroke, patients lose their ability to dream.

Vision is not the only sense that involves different parts of brain reacting to actual or remembered stimuli. Marc Raichle, a neurologist, and his team at Washington University in St. Louis did an experiment in which they either showed subjects a word, read the word out loud to them, or asked them to think about that word. The PET scan images showed that blood flow increased in different parts of the brain depending on whether the word was seen, heard, or thought about. As in Kosslyn's subjects in the shape experiment, when the subjects *saw* a word, the back part of the brain, which integrates visual signals, lit up. When the same word was *heard*, the side part of the brain—the auditory cortex, where sounds are processed—became active. In other words, when presented with the actual sensory stimulus, those parts of the brain concerned with receiving and interpreting the auditory or visual signals become active. But when the subject was asked to think about the word, the parietal lobe lit up. This was the same part of the brain where blood flow increased in Kosslyn's visualizing subjects.

Joe LeDoux, a neurologist at New York University, studying fear pathways in the brain, has mapped out the pathways of hearing-to-fear and vision-to-fear. He found that when we hear a fearful sound or see a fearful object—say, a hissing or rustle and the shiny green streak of a snake in the grass—a tiny part of that sound or image gets shunted to the amygdala, where the brain processes fearful emotions. This happens almost immediately, before the full image or sound is reconstructed. Nerve pathways also lead from the amygdala to the hormonal stress center in the brain, the hypothalamus. So, as soon as we see or hear a threatening object, we experience all the emotions we identify as fear, and as soon as we experience the fear, stress hormones start pouring out, sympathetic adrenaline-like pathways are activated, our fight-or-flight response kicks in, and we are poised to either fight or run.

The positive equivalent of the amygdala—the part of the brain that processes pleasurable stimuli—is called the nucleus accumbens. This pleasure center is located deep within the brain. And although less is known about the brain's reward pathways, it is likely that they have similar connections to sensory and stress centers as do fear pathways. This means that powerful pleasurable signals can also activate the stress hormones and the parts of the brain that control heart rate and breathing. This is the source of the adrenaline rush that comes with sexual excitement—the rapid heart beat, flushing, and breathlessness of the chase. So, through these same pathways, whether we visualize or think about a stressful event, a beloved person, or a hated one, it will be registered in different areas of the brain, but the same hormonal and chemical responses are brought into play as if the actual event had occurred or the person was present in the room. It is through these hormonal responses that our immune responses, and ultimately our health, can be affected by our thoughts and our emotions.

At first it was believed that the direction of such emotions—whether positive or negative—was not as important as their intensity in exerting power over our behaviors and the hormone and nerve responses that those feelings trigger. But more recently, psychologists, such as Richie Davidson at the University of Wisconsin and John Cacioppo while at Ohio State University, have found that subtle emotional charge may trigger different combinations of physiological responses. There are shadings of tone not only in the emotions we recognize but also in the physiological responses linked to them. Different emotions may contain different ratios of those sympathetic nerve, adrenaline-like responses, as well as their counter-responses, the parasympathetic cholinergic responses. And different emotions may trigger different doses of hormones of the hormonal stress response, such as CRH from the hypothalamus, ACTH from the pituitary, and cortisol from the adrenal glands, or different doses of their opposing, soothing hormones—the breast-feeding hormone prolactin and nerve chemicals such as the endorphins and the Valium-like gamma-aminobutyric acid (GABA).

◄○►

From a species point of view, one goal of our emotions is to communicate our feelings to the group so that the other members can react. To that end, words alone are not enough. What, then, does it take to build the social signals that we send? Through what links were John Donne's two lovers so tightly connected? In the minute milieu within the body where nerves abut against nerves, all it takes is a simple molecule—a neurotransmitter—to perpetuate the electrical signal until it reaches its final destination. But come to the end of the line, to the place where one person ends and another begins, and there is no nerve chemical to bridge that gulf. We must rely on social signals transmitted through the senses—through touch, vision, hearing, and smell—to bridge the gulf to tell others how we feel.

Often words express emotions different from those we really feel. In a communication by letter or even by that most recent medium, e-mail, the subtleties of tone are missing for those whose forte isn't poetry. Almost as quickly as the electronic form caught on, crude telegraphic symbols arose. These punctuations fill a need to convey emotion through the electronic screen: a sad face : (, a happy face :), a wink ;). But add just one sense—the auditory system—and one mode of sending signals through that sense—the human voice—and a whole new range of emotions opens up. The same word spoken on the telephone can convey happiness, sadness, anger, boredom. A lowering of timber and intensity of voice can convey closeness, love, or threat, depending on delivery and regardless of the meaning of the spoken word. A rise in pitch or speed, and anger is conveyed. In some languages, such as Chinese, a curse can be conveyed by a subtle change in tone alone. Now add visual cues, as in a movie or on television, and the richness of the communication deepens yet again. With facial expressions and body language, a whole new dimension of communication opens up. And then there is touch. The warm, assuring touch of mother's hand; the lingering feel of a lover's hand in yours; or the quick firm handshake of a colleague or acquaintance— in each instance, fingers against fingers, palm against palm, and yet a multitude of different emotions are conveyed.

One could argue that in the absence of a social world there would be no need to communicate emotions. Sitting by oneself in a dark room requires no need for expressions of emotions of any sort. But when you find yourself in the presence of others, the need for signaling emotions—through words, through voice, through facial expression, through touch—inevitably becomes important, sometimes even necessary and potentially life-saving.

In 1965, Paul Ekman, a psychologist at the University of California in San Francisco, began studying facial expression as a means of communicating emotions. Ekman was one of a long line of scientists to tackle this field over the last 200 years. In 1804, the British physiologist Sir Charles Bell published his treatise, *The Anatomy and Physiology of Expression*. The third edition of this book, published in 1844, inspired Charles Darwin to study the field from his particular point of view. In 1872, Darwin published his own book on the subject, *The Expression of Emotions in Man and Animals*. By then Darwin was able to utilize the most advanced technology of the day, the camera, to bolster his theories with dozens of black-and-white photographs of humans of all ages using different facial expressions. The expressions of animals and anatomical sketches of muscles were still conveyed in lithograph sketches.

By the 1960s, Ekman had moved away from the descriptive anatomy of facial expressions into the reason for their existence. He asked the question: Are facial expressions universal, or are they culture-specific, in expressing emotion? If universal, one could argue that there might be a biological basis for such emotional signaling. Ekman found that some facial expressions are universally recognized, while others are culture-specific. The universally recognized facial expressions are those conveying anger, fear, disgust, sadness, and enjoyment. In fact, each one of these emotions is comprised of a family of emotions—subtle variations in micro-movements of the face that together are universally perceived by an observer as belonging to the same group of feelings. There are 60 different expressions of anger that can be distinguished from a family of expressions of disgust and from another family of expressions of fear. What is striking about this list is that most of the emotions are negative. Of the five, three—anger, fear, disgust—are warning signals of potential danger. Such signals, even in the absence of words, could all make the difference between life and death to members of one's social group: pointing at a poisonous plant with a look of disgust; showing anger to a foe; expressing fear in the face of a threat. All these behaviors help others around to recognize the potential danger you perceive and to which you have already reacted. (Some of these expressions even play a double role of protecting the signaler from the danger. The muscular effect of the disgust expression, for instance, facilitates expulsion of material from the nose and mouth.)

The converse of these distinct negative emotions, all expressed by equally distinct universal signals, are the family of positive emotions that share a single signal—a certain type of smile. This smile is the kind that is infectious, the kind that makes the bearer light up and causes those around to shift their moods. It is perhaps the kind that Norman Cousins, one of the early popular promoters of the power of the mind in healing, meant when he wrote that laughter cures. It is a smile that includes contraction not only of the muscles around the mouth but also of the muscles that circle the eyes, the orbicularis oculi muscles. It is Santa Claus's crinkly grin; it is the

The study of facial expressions as a mean to communicate emotion began perhaps with the British physiologist Sir Charles Bell's 1804 publication of The Anatomy and Physiology of Expression. *A later edition led Charles Darwin to publish his own book on the subject; by then Darwin was able to utilize the most advanced technology of the day, the camera, to bolster his theories with black-and-white photographs using different facial expressions* (right). *This work continues on to this day, particularly in research done by Paul Ekman, a psychologist at the University of California in San Francisco, who has studied the biological basis for this kind of emotional signaling.*

look of pride in a mother's eyes when she watches her daughter graduate from high school; it is the look of exhilaration in your son's eyes when he makes a home run; it is the look of sheer pleasure when one luxuriates in a gently scented bubblebath.

There is a gratification in watching this sort of smile, a gratification in knowing that one might be the object that stimulated it. And there is a motivation to elicit it over and over again. A child will repeat over and over whatever actions seemed to bring it forth in the parent. And, sleep-deprived for weeks on end, exhausted parents of a newborn infant will suddenly experience an unparalleled sense of exhilaration when they recognize that first true smile in their baby. It is a kind of smile that is very difficult to fake. We can consciously produce the lower mouth muscle's grin—but without a great deal of practice, most of us cannot convincingly make our eyes smile. Grin without the eyes, and we produce a distorted caricature of a smile—a smile empty of the positive emotion it is meant to convey. Instead, the signal we send is sinister, even threatening.

Internally elicited emotions may not be as strong triggers of external social signaling—facial cues or body language—as confronting the actual person or object would be. This is not surprising if the biological reason for such cues is to signal others in the social group of impending danger or safety. In the absence of actual threat, there is no need to signal others. But nonetheless, we do unconsciously smile when thinking of a loved one, frown to ourselves when thinking of unpleasant interactions. This may be because thoughts, feelings, and actions are linked both by learning and by instinct. The hardwired pathways that trigger such body language must be very strong.

What is striking about these nonverbal social signals is that they are conveyed not in minutes or seconds but in thousandths of a second. This is what makes it so hard to capture the perfect smile in a photograph. But because expression of emotion comes in a rapid series of fleeting stills, great portrait photographers can capture an emotion in a single click of the camera shutter. These pure universal facial cues are of course not the only social signals we send. There are longer-lived emotional signals that transpire over longer periods, as in a moving picture. These are made up of combinations and sequences of the primary individual emotions, and they convey a whole new set different from the basic set. In a way, the palette that results by combining the basic emotions over an extended time frame, in which sequence and intensity matter, is like the palette of shades that can be created from combinations of primary colors chosen from a color-wheel.

But vision, hearing, and touch are not the only ways that social signals connect us to one another. Volatile molecules, like those we smell, also play a crucial role in social signaling in all species. Mother rats recognize their own pups in part though their odor, and the pungent smell of cloves can mask a pup's specific scent. Thus, a dam presented with a strange pup at the same time as a whiff of cloves no longer recognizes the pup as not her own, and will adopt the stranger. Odor and other molecules floating through air are the one chemical social signal that does bridge the space separating two bodies. In this way, such molecules act like hormones and nerve chemicals that, when released at one site in the body, can affect the functioning of a distant organ. But in this case, the target organ of such molecules is in another being.

Martha McClintock, a psychologist at the University of Chicago, first made this observation in humans when she was a student in the 1970s at Wellesley College. She noticed that at certain times of the month, there was always a shortage of menstrual pads or tampons in the dorm, while at other times there was an excess of these supplies. She began to take a poll of women, asking when their periods started. She found that after living together for some months, the menstrual cycles of all the women in the dorm seemed to synchronize. It turns out that molecules called pheromones, released into the air by women, can play a role in synchronizing sex hormonal cycles, and thus can synchronize the menstrual cycle, too. These gaseous hormones bind to receptors—not odor receptors in the nose, but receptors in an organ inside the nose called the vomeronasal organ. Female pheromones, which have a faintly fishy scent, can act as powerful sexual stimuli in men, and male pheromones, with very little smell, can also stimulate sexual desire in women.

Sometimes signals sent through all these modes—whether visual, hearing, touch, or pheromone—will harmonize, and sometimes they will be out of sync, because these outbound social signals, with the exception of our words, are usually outside our conscious control. Different individuals are more or less adept at sending these complex social signals, and others are more or less adept at decoding the complex signals that they receive. But whether misinterpreted or not, these social signals do convey and do evoke emotions. And the recipient responds with all the network of nerve pathways, neurotransmitters, and hormones that can be activated by other means. Here again the social world can influence our health through the same hormonal and chemical pathways set into motion in other ways. The social world can activate the stress response, or it can tone it down. The effects of these personal connections can be more soothing than an hour of meditation. They can also be as stressful, and more long-lived, as running at top speed for twenty minutes on a treadmill. In fact, of all the sensory signals that impinge on us from moment to moment throughout the day, it is the ones connected in some way to another person that can trigger our emotions most intensely. If emotions are really meant to move us, it is these bonds toward which they push or from which they pull. Whole industries are based on the power of such social bonds: romance novels, movies, cosmetics, fashion, advertising, popular songs. In one way or another, the whole of our popular culture strives toward sealing or healing these social connections.

Once they are made so strong, what happens if we then break these social bonds in childhood, or later on in adult life? The simplest and yet the strongest bond that can be broken is the mother–child bond. Simplest because of number, and yet complex and powerful, it is the template upon which all our other social interactions are based. This core of all social interactions is composed of just one more person than ourself, the pair, or, in social psychological terms, "the dyad": brother–sister, best friends, lovers, husband–wife.

So much does the mother–child relationship color our later life that Sigmund Freud and other developers of late-nineteenth and early twentieth-century psychoanalytic theory attributed entire groups of emotional and physical diseases to disruptions of that bond. Later, in the 1920s, the proponents of psychosomatic medicine—the Hungarian-born psychoanalyst Franz Alexander, working at the Chicago Institute of Psychoanalysis, and Helen Flanders Dunbar, the American physician and theologian at Columbia Presbyterian Hospital in New York—also joined with this thinking. Diseases ranging from gastric ulcers, to heart disease, arthritis, asthma, and depression were all explained in the early twentieth century, with complex circuit-board-like diagrams and charts, as growing from a troubled childhood. Although now the idea of blaming it all on mother seems too pat and rooted in another century when father ruled, early maternal–infant interactions do play an important role in shaping our biological and hormonal stress responses in later life.

Researchers have studied this bond throughout the animal kingdom. A baby rat, for instance, born with its eyes closed and hairless, looks much like a small pink peanut shell, and is just as helpless. It needs warmth and nurture. If separated from mother, the rat pup starts emitting ultrasonic calls, the same as its adult cousins, the bats. In the laboratory, these calls can be detected with a hand-held meter called a bat detector (the same instrument spelunkers use for detecting bats in caves). The longer it is isolated from mother's warm, thick fur, the more the rat pup's temperature falls, and the faster its calls come to a crescendo. A nursing dam will respond immediately to these calls and within less than three minutes will round up all her pups, nudge or carry them to a wood-chip pile she has created in the corner of the cage, and then crouch over them and nurse. The whole is an amazing dance of nurturance and survival, repeated over and over by all mammals with their young. In fact, so rooted in reflex biology is this ritual that even nursing human mothers respond almost instantly to any baby's cry with a milk let-down.

The amazing thing about these responses is that there is a kind of symmetry to the emotions and their underlying hormones in mother and child. It is as if a single social signal—the baby's hungry cry— jumps the gulf of space separating two beings and links their separate physiologies. This social signal, the baby's cry, can be as powerful a stimulus of a mother's stress hormone response as any chemical. The cry, brought on by hunger, and associated with a burst of hormones in the baby, brings on a burst of hormones in the mother. Seeking out the baby and nursing in response to that helpless cry soothes the baby's hunger and in turn its stress hormone responses—the baby falls asleep. But, then, so does the mother, for the same hormone that facilitates nursing, prolactin, is also a calming, soothing drug. Thus a specific social signal in the baby—the hungry cry—elicits a specific behavior in the mother designed to resolve the problem—nursing. And the hormone that permits nursing in turn changes the mother's mood.

Prolactin is not just a hormone that controls milk production in the breasts. It is a hormone that is as powerful an antistress chemical as any drug that has been designed to combat anxiety. This is why nursing mothers may seem oblivious to the stresses and strains of day-to-day hassles, why some seem to have lost their competitive drive in the world and seem so totally focused on the infant they are nurturing. This is why some women yearn for this peaceful state after the baby has been weaned and the cares of the world once again take on their exaggerated proportions. This complementary set of hormonal responses in baby and mother—set up by hunger in the baby, and ending with the mother's sense of peace—all contribute to a powerful bond between mother and child.

But if you break that bond, there will be consequences. From the 1950s through the 1970s, Stanford University psychologist Seymour Levine conducted a series of experiments that found that, in the rat, interrupting the bond daily for only 15 minutes at a time during the pup's crucial first two weeks of life will bring long-lasting changes in the brain's stress response. He found that the pups' blood cortisol rose and stayed high for as long as seven days, after even short separations. And Michael Meaney, a neuroendocrinologist at McGill University in Montreal, later found that even well into adult life the "separated-at-birth" pup will pour out more CRH from its hypothalamus, more ACTH from its pituitary gland, and more corticosterone from its adrenal glands in response to stress. Monkeys repeatedly separated from the mother in infancy are also, like rats, permanently changed. And generations of abusive primate mothers perpetuate the chain in their offspring until it can be interrupted and broken with a nurturing mother.

As it turns out, there are genetic strains of rats that mother differently. In fact, these are the same strains of rats that are either low or high stress responders—the Lewis and Fischer rats. If you separate rat pups from their low-stress-responsive mother, she immediately picks them up, piles them in a nest, and crouches over them—twelve pups quickly coddled under mother in less than five minutes. If you separate pups from their high-stress-responsive mother, she ignores them, steps all over them, and abandons them for more than the usual fifteen-minute observation period. It could be that this difference in mothering contributes in some part to the stress-response differences these rat strains show in later life, just as Meaney's and Levine's separated pups developed increased hormonal stress-responsiveness as adults. Or it could be that the mother's extreme-stress-responsiveness contributes to her abandoning her pups. Most likely there is a vicious cycle in which it is impossible to tell where nature leaves off and nurture begins in this determining of differences in adult stress-responsiveness.

The association between social deprivation and altered physiological responses in human infants was noted as early as the 1950s. In Rochester, New York, the American physicians George Engel and Franz Reichsman described a syndrome of deprivation, which was the opposite of the stress syndrome Hans Selye was popularizing. This syndrome resulted from too little, rather than too much, stimulation, especially during infancy. These physicians developed their theories after many months of careful observations of an infant girl born with a gastric fistula—an inherited condition in which her stomach contents drained to the skin through a hole in the abdomen. While the child was in the hospital, awaiting surgery for repair of the fistula, the doctors and nurses noticed that she became withdrawn and uncommunicative. In the 1950s, hospitals were drab and barren places, especially for children. Isolated from her parents, left alone most of the time on the white sheets in her iron-barred bed, the child became increasingly despondent.

Engel and Reichsman began to monitor the girl closely, attempting to play and interact with her, and at the same time observing her behavior and collecting and measuring the volume and contents of her draining gastric fluid. They found that there were dramatic and immediate changes in these secretions that fluctuated with and corresponded to the girl's moods. When she was relaxed and playful, the stomach juices would flow profusely, but when she was frightened and withdrawn, such as when a stranger approached, the flow of fluid would immediately cease. They theorized that too little stimulation was as powerful an influence on the body as too much.

What do such studies tell us about how much adult stress-responsiveness in humans is related to early childhood events and how much is in our genes? If animal studies predict well, probably less than 50 percent of our stress-responsiveness is in our genes and more than 50 percent is molded by our environment. Thus, although we may be born with a certain set-point to our stress-responsiveness that is preprogrammed in our genes, as in the Lewis and Fischer inbred rats, there is still a large component of that set-point that can be modified by environmental factors, especially early childhood experience. In humans, just as in rats and monkeys, critical time windows in childhood play a disproportionate role in shaping later-life responses to stress. Children abused or abandoned before the age of ten are more prone to later psychological and physical problems when they mature, including increased stress hormone responses and tendencies toward depression. A recent example is found in children of Romanian orphanages. Those children adopted into Canadian families before the age of four months were no different from their Canadian-born peers. But those adopted after the age of eight months showed lasting elevations in stress hormone levels. This suggests that in humans, as in other mammals, a critical time window may exist in infancy where the duration of deprivation, as well as its severity, may play a role in later, lasting physical and psychological effects.

These examples are extremes of deprivation. But what if the far-flung social world we have constructed in this electronic age confers a lesser, but still significant, degree of social deprivation on us all? And what if that degree of social isolation has physiological effects? We don't know the answers to these questions precisely, in part because the powerful effect of the social world on physiological and immune responses is only now being systematically studied.

We take for granted that our children will grow up and grow away, that our grandchildren will live in different cities from ourselves, and that we will change jobs at least a few or perhaps many times in our lives. We also take for granted that with such job changes, we will usually also change the place we live and the friends we have. With all this mobility, we lose our extended families. And then we lose those friends we had found to replace the families left behind. Yet humans are affiliative animals—biologically not meant to spend their lives too far from the pack. We crave affiliation, we seek it—in fantasy, in art, and in all the devices we have invented to overcome the social isolation that our mobile lifestyle generates. So now, stoked by Internet, telephone, and e-mail communications, we have adapted to that lifestyle and have begun to take for granted commuting spouses, and "LDRs"—long-distance romances.

Without surrogate electronic bridges to our social interactions, we would be isolated, completely deprived, able to communicate with loved ones only in the slow-motion pseudo-conversation of a letter. The rapid response time of e-mail and the telephone gives us a sense of true social support. But is this sense real? Or are these pale, uni-dimensional pieces of a social interaction? Peel back the rich layers of a close physical social interaction, peel away the touch, then sight, then voice, down to the electronic words—is this really a social interaction, or is it a soliloquy? What is missing in these interactions is the original meaning of the word "interaction"—the action between people. What is missing is the tennis match of tossing social signals—words and thoughts sent from one person to the other, and responded to and tossed right back in a slightly modified form. The thoughts and words in such an interaction are not static, bouncing back as from a mirrored wall, but changing, growing, shrinking, morphing, as we sense the others' responses. And so such interactions can spawn new feelings, new ideas, new understandings that grow from the relationship and in turn make it grow. It is as if each sense was painted on one of those cellophane transparencies, the "cells" that animators used to make the first animated films. With each new layer, a richer, more three-dimensional form emerges. Add movement and subtle shifts of one translucent layer upon the other, and the dimensionality deepens even more. A face-to-face relationship extended over time has that element of dissolving and re-forming that helps the partners re-meet each other many times from many different starts. Will our relationships, which more and more rely on single modes of social signaling, ultimately suffer from this loss of richness, or will we adapt to these new electronic modes?

Logic suggests that our increasing reliance on electronic communications might increase social isolation. Indeed, some studies suggest that unlimited access to e-mail actually increases people's sense of loneliness and isolation. But there are others working in the new medium of telemedicine who have found that electronic communications may make it easier for people to interact. In isolated communities, too small to afford full-time medical health providers, telemedicine—diagnosis and treatment with the assistance of videoconferencing technology—is being tried out with success. Some who have been instituting telepsychiatric sessions have noticed that rather than being inhibited by this mode of communicating, patients actually feel less inhibited than when speaking to a psychiatrist face to face. This may be part of the same phenomenon as the ease that we feel in communicating our deepest burdens to a respected but impersonal confessor. It may be part of the successful principle behind confessing sins to a hidden priest in the Catholic confessional. However, in this new age, the person to whom we bare our hearts is not hidden behind oak walls and velvet drapes but behind a glowing, protective computer screen.

It may be that the intimacy we can feel in a telephone call also comes from the protective quality of not seeing or being seen by the other person. Another aspect of telephone conversations that may lend them intimacy is the distortion and magnification of a single mode of social signaling—the voice. When two people are face to face, whispering or speaking quietly brings them together, makes them lean forward to hear what is being said. Such moving together physically lends a sense of psychological closeness as well. When we speak on the telephone, there is a sense of being whispered to, which can transmit intimacy. Yet another element of this electronic medium that might lend itself to creating intimacy is a sense of some limitation of time. A clearly delineated time limit, like that of a psychotherapy session, sets a pace that allows a conversation to move quickly from superficial to more sensitive areas, and then back outward to less painful topics. The unspoken demarcation of time in a telephone call can set a rhythm that allows callers to dip carefully in, and then, with relative ease, out again of their more emotionally difficult thoughts.

Thus, such electronic modes may actually help us deepen our relationships and sense of social connectedness, rather than keep us isolated. They can be consciously and judiciously used not only to keep a connection alive but actually to strengthen it between meetings in this world of loved ones living apart: grandchildren talking to grandparents, cousins to cousins, sisters to brothers, husbands to wives, lovers to lovers. The judgment is just now beginning to trickle in on the effects of new electronic modes of communication on our sense of social isolation of connectedness. It may be that in some circumstances they diminish a sense of isolation and in others increase it.

Whichever turns out to be the case, we each must learn to deal with our changing social world, now complicated with yet another means of sending social signals. But no matter how sophisticated the modes of sending signals become, it is still, and will always be, our basic unchanged physiologies that respond. Those early observations of Engel and Reichsman on the effects of deprivation, and those of Selye on the effects of stress on physiological responses, still hold despite all the many years of questioning by the medical community. These researchers first proposed that both these conditions—stress and deprivation, which resulted from either too much or too little stimulation from the external world—were associated with hormonal changes in what came to be known as the stress hormones: ACTH and cortisol. The concept that the body's internal balance could be so profoundly affected by external stimulation, or lack of it, was so revolutionary in the 1930s and 1940s that physicians', and soon the public's, understanding of illness shifted from a focus on the internal to a focus on the external. Many illnesses were soon—and still are—blamed on too much stress, or too little. In the 1950s, the pendulum began to swing away from excessive focus on internal distortions of anatomy as the universal cause of disease to excessive focus on external causes, such as stress and deprivation. And that swing of the pendulum smacked right up against the resistance of the medical community to the popular notion that stress or deprivation could make you sick. This overfocus on one or the other part of the equation—the need to attribute the cause of illness almost exclusively to internal or external origins—probably contributed to the medical community's continued rejection of the role of stress in illness until more recent studies could build a strong case for the operation of these principles in humans. As we saw earlier, one problem that immunologists in particular had in accepting the impact of these factors on immune function was that the time scale of an emotional response is very different from the time scale needed to change a long-lasting immune response such as antibody production.

How long does it take for an image to evoke a memory? Milliseconds. How long does it take for such a memory to evoke a mood? Seconds to minutes. For a mood to trigger hormonal responses? Minutes. And for hormonal responses to affect the workings of our immune cells? Minutes to hours. So a fleeting change in mood is unlikely to have a sustained effect on immune-cell function and on our susceptibility to disease. But repeated, relentless mood shifts, the wearing sort that come from the constant stress of chronic caregiving, a long drawn-out divorce, or painful relationships, are the kind of social stresses that wear us down and make us vulnerable to disease. And they do so, at least in part, through activation of nerve chemicals and hormones that affect immune-cell function.

John Sheridan, an immunologist at Ohio State University, explored disease susceptibility in male mice undergoing repeated social reorganization. He moved the dominant mouse in a cage to another cage and thus a new social group. After a period of acclimatization to its new cagemates, he moved it once again, then repeated the process twice more, for a total of four moves. This sounds like the kind of lives that some executives now live. Each time the mouse arrived in its new home, it fought to reestablish its place in the cagemates' hierarchy. Whenever groups of animals come together, indeed, groups of people too, they quickly establish a social hierarchy. (The phrase we use to describe this phenomenon, a "pecking order," comes from the fact that chickens show their place in social groups by pecking at one another.) There is always a dominant animal in a group—usually a male—who does all the pecking, while the lowest animal on the social ladder is most pecked at. In rats or mice grouped together, this fighting, scratching, and biting can leave the subordinate animal severely scarred. And then there are all those in between the top and bottom.

At the end of the frequent moves, Sheridan checked the mice for susceptibility to herpes viral infection. It turned out that, unlike mice exposed to other kinds of stress—stress that was more physical than social—the mice that had been socially reorganized were much more prone to reactivation of dormant herpes virus. And it was the dominant mice, those who became involved in social conflict more than their cagemates, that developed the worse infection. This was so despite the fact that stress hormones were elevated in all these stressful situations. It seems that social conflict brings out an additional and unique hormonal response that is not stimulated by other forms of stress. This unique pattern of hormonal stress response predisposes socially stressed mice to herpes infection. The hormone that does this, which is secreted in saliva, is called nerve growth factor. Those who are prone to herpes virus "cold sores" will find this situation all too familiar. It is exactly when we are stressed—perhaps with lack of sleep and too much work, but especially with prolonged anxiety over personal or workplace situations—that we invariably get a cold sore.

In fact, Sheldon Cohen, a psychologist at Carnegie Mellon University in Pittsburgh, already had found evidence to support this relationship between stress and vulnerability to viral illness—this time the common cold—in humans in a simple experiment he conducted in the early 1990s. Cohen paid healthy volunteers to be exposed to measured amounts of common cold virus delivered through nose drops in the lab. Before dosing them with the virus, he administered a series of questionnaires to measure the amounts of stress they had been under. Cohen found that compared to nonstressed volunteers, those who received the virus during periods of high stress showed worse cold symptoms and higher viral counts in mucus collected from the nose. The kind of stress that these questionnaires tapped into was the cumulative kind, like those that tip the scale of Bruce McEwen's allostatic load.

In the 1980s Jan Kiecolt-Glaser and Ron Glaser, the psychologist–virologist team at Ohio State University, measured several aspects of immune function in caregivers of Alzheimer's patients. They found these aspects to be decreased both in active chronic caregivers and in grieving spouses months after the loved one had died. These immune differences all fit with the effects of stress hormones suppressing immune-cell function. It is not unusual for a grieving spouse to develop serious illness within months of the loved one's death, and although self-neglect while overwhelmed with caregiving could contribute to such illness, the chronic stress and subsequent hormone increase clearly take their toll.

The Glasers studied another group of subjects experiencing chronic psychological stress: divorcing couples. One of the things the Glasers found is that not all divorcing partners showed the same immune and stress hormone changes. Some were more affected than others. It was the confronting wife, faced with an impervious husband, who showed the greatest increase in stress hormones and greatest decrease in immune cell function. It was these partners who would be more prone to illness in the face of the chronic stress of divorce—illnesses ranging from increased severity of the common cold to increased osteoporosis in premenopausal women. By putting on social armor, disconnecting from and ignoring the problem, the withdrawing spouse is also shielded from the physical effects of social stress. And the sensitive spouse, the one who is not blind to such signs, bears the brunt of this stress.

Another situation in which social interactions with those around us can be stressful occurs in hierarchies. Clemens Kirschbaum and Dirk Hellhammer studied soldiers entering boot camp. In this setting, new recruits are not only exposed to the strenuous activity and new environment of camp but also must find their place in a new social group. Since individuals in a social group can readily and accurately rank themselves and others, Kirschbaum and Hellhammer gave new recruits questionnaires in which they were asked to do just this. During the seven days of boot camp, they measured the recruits' salivary cortisol responses to a psychological and a physical stress: public speaking and mental arithmetic, and a twelve-minute run. Surprisingly, they found that recruits whose ranking in the group was high, the dominant ones, had a much higher rise in cortisol and were, according to their hormones, more stressed, than those who ranked low on the social scale. The explanation for the heightened stress response in the dominant members of this group is related to the responsibility these men felt for life-and-death decisions that they had to make. In other situations, it is the subordinate members of the group whose stress responses are exaggerated. This is true in troops of primates in the wild and in controlled laboratory settings. Subordinate members of a baboon troop have to fight for food and water, and subordinate males in such troops must also fight the dominant members for a mate. In these settings it is the subordinate, not the dominant, members of the hierarchy who show the greatest stress response. Thus standing in a social group, whether dominant or subordinate, can be perceived as stressful and can affect stress hormone responses and susceptibility to disease.

While these examples are taken from the extreme situation of boot camp or experimental situations involving animals, we all experience some degree of hierarchy stress at some point in our lives. Entering freshmen at any level, whether students starting middle school, high school, or college, or adults starting a new job, experience some of the same stresses as boot-camp soldiers. In all these situations the stress of place in hierarchy is compounded by the unfamiliarity of a new environment and the anxiety of inexperience. But one need not be new to a job or situation to experience hierarchy stress. Unless self-employed, we all, at one time or another, experience hierarchy stress in our jobs. If you are a manager, you may feel the same sort of constant pressure to perform at peak that the army recruits felt. You may feel the same responsibility to your troops, and so have a heightened stress response. But if you are lower on the rung, and feel that you cannot control your destiny, you, too, may feel stressed. Add to this cramped, noisy, or exposed physical surroundings—surroundings where one cannot find any respite from the constant strain of work, surroundings that are a constant reminder of one's lower rank—and the strain of hierarchy in the workplace amplifies.

Now imagine a few dozen, a few hundred, or a few thousand biological systems, all with inverted U-shaped curves, all exposed to such unrelenting stresses of a hostile workplace: a competitive, insensitive boss who believes that fear and disparagement motivates; a crowded, noisy, or even dangerous workspace; daily uncertainty of future employment; an inflexible work schedule; lack of support; long hours; no psychological or financial reward for a job well done; a priori assumption that the employee is trying to get away with something. Amplify all this at lower rungs of the hierarchy; amplify this in single mothers who have no flexibility to care for their children when they are sick or out of school; amplify this in women or minorities who repeatedly try, but fail, to break the glass ceiling. You end up with a few dozen or a few hundred or a few thousand more or less synchronized inverted U-shaped curves, most of which will be on their descending limb. Integrate all of these into a single whole, and you describe the whole workplace's inverted U-shaped curve. Surely such a workplace is neither biologically healthy for its workers nor productive. In this sense, constructing an atmosphere at work that minimizes such stresses should not only help workers cope but also optimize productivity, benefiting all.

All these examples suggest that difficult social interactions can have negative effects—can act as stressors to stimulate stress hormone responses and thus increase the risk of illness. But being embedded in a social group need not be negative. In fact, recent research tells us that a very important aspect of maintaining health is having an extended network of social support. Sheldon Cohen's most recent studies on stress and susceptibility to the common cold, conducted with immunologist Bruce Rabin at Carnegie Mellon University, showed the surprising result that people who had more different kinds of social interactions in a day had a lower, not a higher, incidence of illness. This finding was not intuitively obvious, since one might predict that someone with more contacts might be more likely to be exposed to a virus.

In the early 1990s, David Spiegel, a psychiatrist at Stanford University, and Fawzy Fawzy, a psychiatrist at the University of California in Los Angeles, found similar positive effects of social support on cancer outcome. Fawzy studied men and women with malignant melanoma, and Spiegel studied women with breast cancer. Both showed that psychiatric interventions such as group therapy and training in coping strategies prolonged life when used in conjunction with cancer drug therapies. While leading a social support group for women with breast cancer, Spiegel found that during the months of group therapy, the women formed deep emotional bonds with one another. With Spiegel's and his staff's guidance, they learned to express their emotions—their sadness, their anger at having cancer. Many eventually achieved a sort of peace—an acceptance that their cancer was a part of them and yet separate from them. Outside the groups, if times were tough, the women called each other for emotional support. The ways that women in these groups lent their support included the same ways we use to signal any social interaction through all of the senses—a tone of voice, a touch, a sympathetic glance. The women who benefited less were the ones who maintained their distance, maintained their wall around themselves.

When Spiegel looked at the life spans of these women and compared them to those of women who were not part of a group, he found that the women with the strong social support network lived longer—many months longer—than their peers. In such a complex setting, where different women are receiving different drugs for cancer treatment, where some may work through their conflicts more and others less successfully, it is hard to identify the ingredient in the groups that prolongs life. But it is certainly possible, particularly in light of everything else we've learned about stress, that the women who developed successful coping mechanisms also learned to tone down their stress hormone responses in part through the social bonds they formed. Indeed, John Cacioppo, when at Ohio State University, showed that gestures of social support from a significant other—as simple as hand-holding—can reduce physiological and hormonal responses to a stressful stimulus. If this is the case, then one could imagine that with fewer bursts of steroids to pick off cancer-fighting and infection-fighting immune cells, the body's defenses, and thus life, could be prolonged.

Just as there is good stress/bad stress, there are positive and negative effects of our social world on health. Too little interaction—loneliness—and we can wither; too much negative social interaction, and our stress response goes into overdrive. But a rich and varied fabric of positive relationships can be the strongest net to save us in our times of deepest need.

CHAPTER · 9 ·

Can Believing Make You Well?

————◄o►————

If stress can make you sick, can believing make you well? And if we just work hard enough at it, shouldn't we just be able to think ourselves better? Logic and perhaps human nature tell us that the answers to these questions must be yes. In the 1920s the French psychoanalyst Emile Coué wrote that if one looked in the mirror every morning and recited the lines "Day by day, in every way, I am getting better and better," one would actually get better. In the 1970s, Norman Cousins espoused laughter and a positive attitude as a healing therapy. Thousands of fervent followers of these prescriptions for well-being did get better, and if their illnesses weren't resolved through these techniques alone, at least they gained the strength to fight it through. We speak of patients fighting illness, battling cancer—all terms implying active and conscious participation of the patient in the fight against disease. In fact, no physician who has dealt with dying patients would deny the power of the will to live. Fight one more month until the grandchild is born, one more week until the sister returns to say good-bye, and the patient lives; but once the will is lost, the fight is over, and the patient soon slips away.

But the question still remains: Can we consciously choose to improve our health? The answer to this question lies in knowing what portion of the systems that control our health are hardwired and unchangeable, and what portions can be changed by how we think and what we believe. Believing is many things. It can be fervent prayer. It can be thoughtful meditation. It can be deep conviction. Or it can be a set of assumptions so ingrained that we don't even realize they're there. One element common to all these forms of belief is expectation—we pray, we laugh, we repeat a phrase, we take a pill, and we expect that these actions will help us heal. And at the core of such expectation is learning.

Just as we can learn a new task, we can learn to make connections among events, actions, and feelings. This form of learning, called conditioning, becomes automatic, and it comes from the brain's ability to associate two signals. It is the classic conditioning experiment, in which the Russian scientist Ivan Pavlov trained his dog to associate the sound of a bell with dinner so that whenever the bell would sound, the dog would salivate. A psychological stimulus is paired with a physical one, and each can then trigger a physiological response.

We all carry with us associations, some good, some bad, that have been learned through such repeated pairings. Step out of the elevator at 8:00 A.M. into a hostile workplace, and you experience all the flood of bad feelings that have battered you each day, even before your boss confronts you yet again. With such feelings come physiological responses: your heart rate increases, you sweat, your stomach turns. If the anxiety is great enough, your blood pressure may increase—these are all of the classic fight-or-flight physiological responses. But step out of a cold winter night into your health spa, smell a warm whiff of swimming pool chlorine, and instead your muscles relax, you feel a glow with the anticipation of the eucalyptus-scented sauna and whirlpool.

These feelings, too, are learned by association. We are not born with a set of beliefs that make us respond negatively to the workplace or positively to the health spa. We learn them after repeated pairings of the stimuli. A workplace can elicit a positive set of physiological responses if the environment is supportive, nurturing, rewarding, instead of hostile and unsupportive. A health spa can elicit negative associations if you had an accident there, feel ashamed of your looks, or can't accomplish your goals. The common factor that can affect your health in these situations is not the physical or psychological stimulus, but your body's physiological response. And such learned associations change the body's nerve and hormone responses, which ultimately affect how immune cells work. In that sense, believing *can* make you sick or make you well.

Perhaps if we could relearn a new set of associations, turn negative into positive, we could in some sense consciously control our health. Perhaps with practice, we could learn to disconnect the feelings from the events that bring them on—through conscious will, through psychotherapy, through meditation or prayer. It then takes one more step to imagine that the emotions that come attached or disconnected could trigger the nerve and hormone pathways that could change the immune system and thus our physical health.

In the 1970s, and even today, the idea that the immune system could be taught was considered an outrageously heretical notion by most classical immunologists. Immune cells can't think; they can't be trained, like dogs, to do tricks. Immune cells respond to molecules that crash up against them, get stuck to proteins protruding on their surfaces, get gobbled up into the cell's interior, and cause other molecules to be made in the cell's protein factories and spit out—magic antibody bullets that surround the prey and destroy it. There is no room for learning here.

But in the 1970s, an American psychologist at the University of Rochester, Bob Ader, and his immunologist colleague Nick Cohen, decided to test whether the immune system could be trained, like Pavlov's dog, to respond to a conditioned stimulus. They paired the sweet taste of saccharine with an anticancer drug that suppresses immune function, cyclophosphamide. They fed both the drug and the saccharine to mice over and over again. Each time the immunosuppressive drug was given, as would be expected in response to such a drug, immune-cell counts went down. Then Ader and Cohen took away the drug and fed the mice saccharine alone. The immune-cell counts fell again, this time in response to the saccharine alone. The saccharine had no intrinsic chemical ability to lower the number of immune cells—before the learning took place, it had had no effect at all. The mice weren't telling their immune cells to fall, weren't thinking "Let's turn off these cells." Something automatic was happening, something that had been learned. The mice were associating the sweet taste of saccharine with the immunosuppressive drug, and after making the association, all it took was the taste of sweetness to affect the cell counts.

Ader and Cohen did this work before much was known about the molecular neurobiology of learning and before the notion was widely accepted that nerve chemicals or hormones could affect immune-cell function in a physiological way. Their report in the 1980s that conditioning could actually change the immune response, and might therefore explain some part of the placebo effect, was met with derision from some scientists. Even some of the scientists who themselves were coming at the brain–immune connection from a different angle—from the stress hormone and nerve pathway route—were skeptical. In fact, although they could not yet admit it then, the work of both sides was converging, and proving the same thing. Those whose studies had started at the conditioning end had concluded that this part of brain function—learning—changed the way the immune cells fight disease and could possibly even partly substitute for immune active drugs. Those who had started with the chemicals and hormones of the brain had concluded that these compounds changed the way immune cells work, and vice versa, that immune chemicals and hormones changed the way that brain responds. What was missing then to bring the two sides together was the detailed knowledge of the chemicals and nerve pathways of learning and the discovery that some of the molecules of learning are those same immune molecules, the interleukins, that make immune cells grow, divide, and mature.

What is learning, if not another set of cells growing and maturing to make connections and spit out another set of chemicals in a different organ—the brain? It turns out that nerve chemicals such as serotonin can make nerve cells grow connections. Drip a fraction of a drop of serotonin on a nerve cell in culture, and you can watch the sprouting bud through time-lapse photography through the microscope objective. At first a little hump appears; and then a point and finally an elongating filament emerge from the place on the cell membrane where the serotonin touched, as if some invisible hand has pulled it out like taffy. Does this kind of sprouting play a role in learning? Are we, as we learn, actually growing new nerve connections?

We do a task repeatedly, shakily at first: once, twice, fifty times, and suddenly, as when we learn to ride a bicycle, play the piano, or type, it clicks and the task becomes routine. Something happens to our brains and nerve cells in this process that changes them forever. On average it takes about fifty repetitions for the change to gel. And it helps to repeat the task before sleep—sleeping helps the whole thing set. There is a biological reason that your mother told you to practice the piano every day, that your teacher made you write out lessons fifty times, and then to get lots of sleep besides.

It used to be thought that the adult brain did not change, that our brain cells, unlike childrens', did not grow and divide and were not malleable. But Eric Kandel, a neurobiologist at Columbia University, found otherwise. Kandel studied the lowly aplysia, a sea slug whose nervous system consists of just the simplest circuitry—and yet it learns. He discovered that nerve cells actually grow new processes, form new connections during learning. Aplysia learn in a very simple sort of way: Repeated electrical stimulation of these simple organisms' nerves, the sort that occurs in higher mammals during learning tasks, produces a permanent change in the aplysia's nerve cells, a change in which new proteins are incorporated into those cells. And with these new proteins, new cell membrane is woven that helps them bud and sprout new connections. We don't yet have proof that this same budding growth occurs in human learning, but more and more studies show that the adult human brain does change with learning. Kandel, in his lectures to students, sometimes used to bring home the point by saying that when you leave this room, if you have listened hard, and learned, you will most likely leave with a few more nerve-cell connections than you had before you came.

In higher order mammals, there are rapidly dividing cells deep within the brain, in memory centers such as the hippocampus. Such cells move up through layers of other cells, to settle eventually in the calmer, more anchored areas of the brain. Such movements and cell divisions happen more with learning. The newer connections that form need to be tended, fired up repeatedly until they are firm, by constant repetition of the task, until it is so deeply entrenched in memory that it becomes automatic. We don't consciously tell our brain cells to make new connections during this learning process, we don't will our cells to grow and divide, but instinctively we know what to do to keep them fired up and activated: we practice the task.

The long-lived electrical discharge that occurs in nerve cells during learning is called "long-term potentiation," or "LTP." It turns out that an immune molecule, interleukin-1, made by those scaffolding cells in the brain, or maybe even by nerve cells themselves, can affect

"LTP" and LTP in turn can make these cells produce IL-1. Here again is an example of an immune molecule playing a direct role in a brain function—a reciprocal role, where immune molecules can change learning and learning can alter immune function.

There is an element of this sort of learning in every prescription we take: we have learned that medicines can make us better. We believe it. The amount of actual improvement in illness that comes from this learned expectation is called the placebo effect. It is the psychological component of that cure. About one-third of the therapeutic effect of every pill comes from the placebo effect. Strictly defined, a placebo is any sort of "inert" treatment whose therapeutic effect is not specific for the disease or symptom it is used to treat. A sugar pill of the same shape and color as a prescription drug can be used as a placebo. Today placebos are used in medical research to determine what part of a medication's effect is specific to the drug and what part of the benefit is psychological. In the first half of the twentieth century, physicians recognized that the placebo effect was a powerful healer, and they used placebo sugar pills to treat illness, not just to test a drug's effects. In 1946, Eugene DuBois, a physician and professor at Cornell University, alluded to the power of merely receiving a prescription when he wrote:

> You cannot write a prescription without the element of placebo.
> A prayer to Jupiter starts the prescription. It carries weight, the
> weight of two or three thousand years of medicine.

—◄o►—

Walking on the marble flagstones in the ruins of the temple to Asclepius, you feel the exhausting heat and blinding light of the full midday sun. But when Lentas was Lebena, a bustling port town in 500 B.C., the healers knew the power of the temple cures. It was then a quiet sanctuary far away from the city's heat and noise below. The earth was not dry and parched. A cooling spring coursed through shaded sage and mountain laurel toward the inner courtyard of the complex, just below the patients' rooms—stone chambers built into the hillside. And the now exposed marble flagstones of the temple, hot from the sun, formed the cool marble floor of the altar's dark interior.

On arriving at the sanctuary, the ailing patient first bathed in the cooling waters of the spring, then entered the dark halls of the enclosed Asclepion and made an offering of honeycake at a small alter to the gods. Later on at night, the patient was ushered into the silent interior chamber, where a priest guided him to lie down upon a pallet on the floor. There the patient slept and dreamed. Asclepius visited the patients in their dreams and, depending on the problem, tended ills with unguents and lotions and cooling fresh water, prescribed nutritious diet and exercise regimens—walking, bathing in the sea— or interpreted dreams. In the Asclepion in Corinth, patients could stroll quietly along a shaded colonnade of an inner courtyard, where the sunlight filtered to the center court. For the lame, a long and gently sloping ramp gave easy access to the temple and the colonnade. Adjacent to the courtyard were dining rooms, where healthful food was cooked and served. Fountains and pools, couches and tables provided soothing spaces for rest and recuperation. The wealthy and the poor could participate equally, as wealthy patients were expected to pay accordingly, and those who shirked were penalized.

Patients came to these sanctuaries, prayed and rested, bathed in the cleansing waters, slept and strolled, ate nutritious meals and were treated with potions and plasters and mixtures made from mountain herbs. Many were healed and left carved stone or clay representations of the body parts that had been healed—legs, arms, hands, fingers, genitals, breasts, heads, ears, eyes—as anatomical votives of their gratitude to the healing god. Such body parts could be conveniently purchased, ready-made, in shops near the Asclepion. These clay statues and their accompanying inscriptions now make up an amazing inventory of testimonials to the cures of these first hospital/health spas.

Some of the healing power of these sanctuaries could have come from the medicinal cures the priests applied. A potion called theriac, which was used from Roman times until World War II, and possibly in various forms before, was concocted in the second century A.D. by Galen, the physician to the Roman emperor Marcus Aurelius. It contained many ingredients we know to be powerful drugs. Proscillaridin, a cardiac drug used widely today, comes from one of the medicinal herbs used in theriac: squill, or *Drimia maritima*, a member of the Liliaceae family. Another, rapeseed, is a source of plant estrogens, and a third constituent, frog skin, we now know contains antibiotic chemicals called maganins. Also in theriac was the seed of a species of poppy, *Papaver somniferum*—the source of opium. Similar plants containing healing chemicals were used by the ancient Greeks. When applied appropriately by a trained physician/priest, they could well have had medicinal healing effects; although in the wrong circumstance and dose, they could also have harmed.

Purifying water was an essential component of all Asclepions' ritual cures, from the first bath before entering the sanctuary to prescriptions for drinking from or bathing in the sacred springs. The cleanliness and cooling, as well as the regular exercise from bathing in the sea, could all have helped effect a cure. Diet and nutrition, too, were essential elements of the Asclepions' prescriptions and certainly contributed to cures. At least some of their effect must have come from the peaceful rest away from city heat and dirt and stress. After observing the ruins of temples to Asclepius in Corinth, in Epidaurus, and at Lebena, Plutarch, writing in the first century A.D., observed:

> Why is the shrine of Asclepius outside the city? Is it that they considered it more healthful to spend their time outside the city than within its walls? In fact the Greeks, quite reasonably, have their shrines of Asclepius situated in places which are both clean and high.

Some more recent historians dispute the notion that most of the beneficial effect of the Asclepions came from their health-spa-like atmosphere, however, since not all Asclepions were located in such refreshing surroundings. Some lay low in swampy, mosquito-infested ground, others in heat not far from the city center. So what was the basis of the temple cures?

The temple healers flourished alongside the ancient world's burgeoning center of the scientific practice of medicine. The ancient Greek physicians, thought by some historians to be itinerant craftsmen at this time, developed skills in prognostication in order to maintain their competitive edge among clients. Through thoughtful listening to patients' complaints and careful observation of their physical signs, these physician-craftsmen noticed patterns of disease that could be used to predict its course. And since the best prognosticators attracted the most clients, the more sensitive the inventory of signs and symptoms that could be compiled, the more successful the physician.

From such practical beginnings, these Greek physicians' approach became the foundation on which modern medical diagnosis is based. A patient's history and the physical, when performed thoroughly, are still the physician's most powerful diagnostic and prognostic tools.

The major ingredient that set apart the temple healers from these practical physicians and their developing science was the patients' belief that prayer and the gods would heal. The other components of the cures—diet, the balance of exercise and rest, the herbal cures—were all also practiced by physician-craftsmen of the day, not gods but mortals like Hippocrates. It was fervent prayer and expectation that brought the sick to these shrines, often with the blessing of their physicians, who, when mortal cures had failed, recognized the immense power of prayer. It must have been the power of belief that forged the temple cures: the ultimate placebo. To say this is not to denigrate their effect— the placebo is a very potent cure, since at least one-third of the effect of any cure, whether modern medication or health regimen of any sort, comes from the belief that it will cure, from the placebo effect.

◄o►

The separation between scientific medicine and the parallel practice of priestly cures began with the cult to Asclepius, continued through early Christian times up through the Renaissance, and, with an ever-widening gulf, to our modern day. In Padua, the university's botanical garden is just across the cobblestone square from the cathedral to St. Anthony, the patron saint of healing. The university's anatomical dissecting theater is a few short blocks away. While Renaissance scientists classified and planted medicinal herbs in that garden, and defined anatomy in the dissecting theater, pilgrims came in droves to the cathedral to pray to St. Anthony. Great Renaissance artists and sculptors adorned his shrine. The healed left tokens of their gratitude, small silver hearts and limbs in thanks—echoes of the anatomical votives left by ancient Greeks to Asclepius.

In nineteenth-century Lourdes, miraculous cures were effected through the intercession of the Virgin Mary, who had appeared there to St. Bernadette, as if in a dream. And in Montreal, at the turn of the twentieth century, a great gray limestone and now-weathered green copper domed shrine, St. Joseph's Oratory, was built to Brother

Just a few steps away from the medicinal garden of the University of Padua's botanists was the great Basilica San Antonio, dedicated to the patron saint of healing, Saint Anthony of Padua. Four hundred years ago these two approaches to healing—scientific and spiritual—flourished side by side.

André, a humble priest who cured with the laying on of hands. The sick still labor up hundreds of stone steps to reach the shrine, on the northwestern slopes of Mount Royal, to enter and pray. They leave their crutches behind as testimonials to the healing power of their saint: more echoes of the votives of the past.

Scientists cannot explain these cures, and with a failure to understand often come skepticism and scorn. But if the belief in the power of a drug to cure gives it at least a third of its ability to heal, why shouldn't prayer and fervent belief alone be an effective cure that works in part through those same nerve pathways and hormones that transduce the placebo effect?

There are two sides to prayer—one is what happens to the believer praying, and the other is the prayer itself. Science cannot presume to have an explanation, nor to refute or to support the intrinsic power of a prayer. But we can now address the first part of the equation—the things that change in the person praying. If prayers do heal, and they surely do, at least a part of their effect must be placebo: the belief that they will heal. To say that the part of healing brought on by the act of praying could come through the placebo effect, is not to say it is fake, but rather to give it a very real explanation. However the placebo effect is brought into action, whether by making a prayer or by believing in a pill, once in play, it acts through well-defined nerve pathways and molecules—molecules that can have profound effects on how immune cells function. A part of prayer's effect might come from removing stress—reversing that burst of hormones that can suppress immune cell function.

Recall the ways that novelty brings stress—through loss of something old and anxiety over the unfamiliar new. As newness fades, so, too, does stress and all the hormone bursts that can exacerbate illness. One way we soothe the stress of change and loss is with old, familiar ritual. Think, for example, of the calming warmth of sitting by a fire. There is a ritual in building a fire in the hearth, a ritual that carries with it its own warm and soothing space, even amidst the tumult of the unfamiliar. The systematic action of piling logs on kindling, the smells of burning wood, the mesmerizing flames, the warmth, are all familiar and bring a sense of peace. All these sensory inputs—the flickering light, the hickory smoke, the cold outdoors and crunchy snow—may hark back to childhood holidays and a simpler time of safety and less responsibility. If we focus on the fire, we can block out the now threatening world around us.

But soothing rituals needn't only re-create the physical. We can re-create a soothing familiar space in our mind through thought and words and song. These rituals are even more portable than such actions as building a fire. These are the rituals of prayer. We can carry them within us and pull them out at any time and any place— at work, at home, in a house of God. We just need close our eyes and we are there, back in that old familiar soothing place that banishes anxiety and grief and sadness with familiarity.

At least one element of prayer that helps remove stress and helps impart a sense of peace is the repeated performance of a set of actions until they become automatic. In the automaticity of ritual, a kind of familiarity is reached through learning. It is as if the ritual is the door through which the believer leaves the hurly-burly world behind and steps into a calming space. Our need for such escapes is present all our lives and reflected in the stories we grew up on. The magic talismans of children's literature all help in these escapes: slip on the magic rings in C.S. Lewis's *The Lion, the Witch and the Wardrobe*, and you can jump into the dark magic pools in the wood between the worlds and enter new worlds. But lose the ring, and instead you land in mud. At times of stress and turmoil, there is a longing of the human spirit to step as if through magic, into such a pool or through such doors, into a child's world of nurturance and peace.

In fact, often the believer does step into a familiar physical space. The ancient Greeks stepped into their temples; later, Christians walked into awe inspiring Gothic cathedrals; modern worshipers go to the temples, churches, and mosques of their own particular faith. In all this, the worshiper leaves an overstimulating and often unfamiliar physical and mental world for a time, to step into the calming familiar place of prayer, wherever that may be. Because even if the worshiper is in novel surroundings—the person who has moved homes, is in a new city, is in a new psychological place of divorce or grieving—there is a constancy to the temple, church, or mosque, and a constancy to the ritual and the prayer, that gives familiarity. With familiarity comes resolution of anxiety. And with familiarity also comes a lowering of the stress hormones that peaked during novelty. So it may be that the familiarity of ritual and of prayer affords the same dampening of the stress response and of anxiety as does becoming familiar with a novel space. If that's the case, then soothing prayer may soothe immune responses, too—remove them from the viselike grip of steroid hormones that suppress their function.

Indeed, although the mechanisms still need to be proven, researchers such as Harold Koenig of Duke University are now finding in large epidemiological studies (of close to 1000 people) that religious observance is associated with less medical illness and lower rates of hospital admissions. On a smaller scale, for more than a decade AIDS researchers have shown an effect of belief and expectations on course and outcome of disease. Neil Schneiderman, a psychologist at the University of Miami in Coral Gables, Florida, found that stress-management training in HIV-positive men buffered not only their feelings of stress but also their cortisol and other stress hormone responses. Margaret Kemeny and her group at the University of California in Los Angeles studied coping patterns in bereaved HIV-positive men. They found that those who processed their loss and found meaning in it had less rapid decreases in number of helper, or CD-4, lymphocytes (those lymphocytes that rapidly fall with progression of AIDS) and lower death rates from AIDS than men who didn't have such positive response patterns to stressful events. This same group studied non-AIDS subjects and also found an association among mood, expectancy, and immune-cell responses. The more optimistic the person, the less an event was perceived as stressful, the more robust were their immune-cell responses.

Putting all these studies together with what we know of the suppressive effects of stress hormones on immune responses, one could infer that the healing effect of belief and expectation might come only through the removal of stress and the reduction of the immune-suppressing hormonal burst. But could there also be a positive effect of prayer, one in which the addition of other soothing molecules plays a role in healing as much as the subtraction of stressful ones? Herbert Benson, a professor of psychiatry at Harvard University School of Medicine, has proposed that there is indeed such a positive response that can be brought into play by all sorts of soothing actions, including prayer and meditation. This "relaxation response" is a stereotypical physiological response made up of a cascade of nerve chemicals and hormones, a sort of mirror-image of Selye's stress response.

◄o►

Across Mount Royal from St. Joseph's Oratory, a pleasant walk through landscaped park and past some wilder woods, is McGill University's medical school building. A round cylinder of glass and steel, the building perches on the southeastern face of the largest of Mount Royal's several extinct volcanic domes, overlooking Montreal's downtown skyline and the St. Lawrence river. In the 1970s, in his introductory psychology class to the medical students, Ron Melzack, a psychologist and expert in pain pathways, described injured soldiers at the Anzio beachhead in World War II. These men, pulled from the carnage, severely injured, missing limbs, nonetheless felt no pain. For them the war was over. They were going home. Loss of limb meant they were saved, and somehow this strong positive psychological meaning of the event overrode the physical component of the pain.

When a limb is cut, electrical signals race from the tiny nerve endings in the skin and muscle, up through those nerve trunks the anatomists discovered, back to the spinal cord. There they enter the spinal cord at bulbous switching stations, tangles of nerve processes closely apposed on nerve-cell bodies whose long processes cross the spinal cord to their cable paths on the opposite side. From here these wirelike nerve bundles thread up the spinal cord to another switching station at the base of the brain, and from there another relay set of nerve fibers leads on into the an area of the brain just above the hypothalamus, called the thalamus. This structure deep within the brain shows the same geographic "somatotopic" distribution of nerve cells as the sensory map that Penfield and Milner described. In this way, when the skin is cut, you immediately know where that cut is on the body, and you feel pain. All this happens in fractions of a second.

But there are also descending nerve pathways, cable paths in which electrical signals move downward through the spinal cord. The sources of the electrical spikes that move along these trunks are the parts of the brain that govern thought and those amorphous parts that we are just beginning to understand, the centers for emotion. These centers and the signals that they send can block the pain signals that arise at sites of injury and then move up the spinal cord. So, the circuitry of pain includes not only a sensory element but also wiring that can dampen pain through psychological tinge.

The pain of childbirth is said to be one of the ten worst pains in medicine. And yet a woman can be taught in part to control that pain. The pain is there; it doesn't go away. But one can be distracted from it by rhythmic breathing and concentration. At least as important as the mantras and the rituals learned in birthing classes is the soothing presence of the guiding nurse or spouse. This psychological support numbs the component of the pain that comes from fear. No matter how knowledgeable the mother is of anatomy and the physiological process of childbirth, at the moment of an intense contraction, sheer terror takes over, in addition to the pain. Because the sensations, never felt before, are so unnatural, to your conscious brain it feels as if your insides are being ripped out. Knowing that this is not the case, being reminded of it by a trusted person, dissipates the fearful part of pain. Learning and association—conditioning—in this setting can help relieve some pain.

The way this is achieved is through a complex wiring system of nerves connecting the brain's higher centers to the spinal cord. It is here that these nerves meet other nerves that send pain signals from the limbs and organs down below. If you crack the spinal column, there, between its hard bony arches and the vertebral column's thick supporting cylinders of bone, lies the glistening spinal cord. Cut horizontally through the spinal cord, and you see a butterflylike imprint: a gray center surrounded by a white outer layer. The soft gray spongy part is where the nerve-cell bodies lie, and in the whiter areas are the cable trunks—the wirelike fibers of each nerve. The butterfly's horn-shaped parts, called "dorsal horns" because of their shape and the back (or dorsal) part of the column where they lie, form a kind of railroad switching yard where tracks from many parts converge. Here the long nerves running up from skin and muscle meet with other long nerves running on to the brain. Shorter nerves, so short that their fibers don't extend outside the dorsal horns, act like switches on the railroad track—switches that can shut off one track and activate another.

Electrical impulses from the higher learning centers travel down the spinal cord and end on other nerves whose Morse code of electrical discharges, moving up the spinal cord, would signal pain. These downward signals can shift the track and thus block the pain. But switches in the spinal cord are not mechanical—the nervous system relies on chemicals, called neuropeptides or neurotransmitters, to do the job. These chemicals released from nerves whose electrical signals are heading down descending pain pathways, or those released from the short inhibitory neurons in the dorsal horns, can dampen pain and block incoming pain signals by temporarily paralyzing the electrical activity of target cells. Among the chemicals that can do this are the endogenous opiates, or endorphins.

—◄O►—

If you face the sea, from the hill on which the temple to Asclepius stands in Lentas, you look south over the shining Mediterranean towards Egypt. But if you turn and face the north, you will look out over the fertile valley that fills the center of the island. On the other side of this valley lie the ruins of the most powerful city of the Minoan

From the sticky white sap of the opium poppy (Papaver somniferum) comes the black gumlike mixture of chemicals that have been used for thousands of years to induce trancelike sleep and silence pain. These chemicals—codeine and morphine, among others—work their magic by mimicking the body's own painkillers, the endogenous opiates. Made in many parts of the brain, these include enkephalins, soothing molecules squirted out by the same cells in the hypothalamus that make the stress hormone CRH. And the same stresses that release CRH and other hormones also release these numbing, lulling molecules.

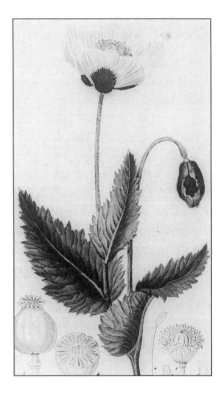

civilization, which flourished around 1900–1500 B.C. It is the Palace of Knossos, home of King Minos and the myth of the Minotaur. Amongst the ruins found in Knossos, dating from about 3500 B.C., is a terra-cotta head—an image of a sleeping goddess, with a beatific smile and closed eyes. On her head is a crown of three slit poppy pods.

There is little doubt that the opium poppy was used—indeed cultivated—by the ancients who inhabited the Mediterranean basin. It is equally clear that these civilizations knew of its medicinal and altering effects. Even the Sumerians, who inhabited Mesopotamia around 3400 B.C., used two written symbols for the poppy: "joy plant."

If, after it has bloomed, you slit the ripe pod of the opium poppy with a sharp knife, a white sap oozes from it. It looks as if someone has tried to repair the pod with too much children's school glue. In the sun, the sap dries down to a black gum containing a mixture of chemicals that can induce a trancelike sleep and silence pain. Multiringed structures made up of carbon and hydrogen atoms, these chemicals are called opiate alkaloids and include codeine and morphine. Extracted, purified, and synthesized, many of them are used in medicine today for pain and for sedation. And they also are, and have been for centuries, drugs of abuse.

The reason these drugs work is that they mimic molecules made by nerve cells, the neuropeptides and neurotransmitters. Each different opiate molecule fits neatly into a protein seat that protrudes from the nerve cells' surface. Once the drug has settled in its cove, it trips another protein latched onto the receptor's root, dangling inside the cell. This then triggers a cascade of protein shifts that lead to changes in the cell's electrical activity. The opiate drugs work because their three-dimensional shape is identical to the shape of parts of the endorphins, which are meant to lock into these receptors. So, while we tend to think only of external drugs as having real painkilling power, the body's own endorphins can shut down pain just as easily and as well as morphine.

Opiates produce their effects on pain in many ways. When released from nerves within the dorsal horns, they can block the electrical activity of pain nerve fibers sending signals up the spinal cord. When released from higher centers, they can increase the firing rate of nerve fibers whose electrical spikes run down the spinal cord, and thus block the upward traffic. In a center at the base of the brain, just above where the spinal cord widens to a stem, in an area called the brainstem, there are two kinds of cells called "on" and "off" cells. These cells can switch on pain paths, or they can switch them off. Morphine and other opiates make "off" cells work harder. (The antimorphine drug naloxone turns "on" cells on and "off" cells off.) Even higher brain centers can also be shut down by opiate drugs and endogenous endorphins. These include the hypothalamus, the amygdala, and higher regions that govern thought. It is by blocking nerve cells in these parts of the brain that opiates cloud thinking, dull anxiety—and also dull pain.

Where do the endogenous opiates come from? They come from nerve cells in many parts of the brain. These cells produce the molecules that soothe us, such as the enkephalins, short strings of amino acids with opiatelike effects. In yet another perfect symmetry, they are made and squirted out by the same cells in the hypothalamus that make the stress hormone CRH. Other opiate peptides are made in cells within the brainstem and spinal cord. The same kinds of things that make the hypothalamic cells release their stress hormones also make them release these numbing, lulling molecules. It is as if by grand design the molecules that oppose stress are there to keep the whole in balance. It is not hard to imagine, then, how such nerve circuits and neurotransmitters can explain the effects of placebo on pain. And if such learning and expectation are part of what we call belief, why couldn't believing change our sense of pain? And if this is the case, why couldn't learned and conditioned beliefs also change the course of inflammation through the same molecules and nerve routes? Indeed, it turns out that blocking the body's own opiate molecules with drugs can block those immune-suppressive effects of conditioning that Bob Ader showed in mice.

Surgically blocking spinal pain pathways can also change the course of inflammation. Two rheumatologists from California, Jon Levine and Paul Green, took advantage of the body's symmetry in nerves and joints to test just that. They cut the nerves on one side in rats, then treated the animals with an oil that caused arthritis. While arthritis developed in the joints on the side where the spinal cord had not been touched, none developed on the side that had been cut. In this case, it seemed, something in the intact nerves was enhancing inflammation. This fit with clinical observations in people who had suffered a paralyzing stroke. If arthritis develops in such a patient, it affects only the joints on the side that is not paralyzed. So these patients and what happened to the rats tell us that something in the nerve pathways in the spinal cord allows arthritis to develop, and blocking these routes stops arthritis in its track. That something could very well be the very molecules released by nerves springing from the spinal cord and feeding those areas of inflammation in the joints and tissues.

―◄◊►―

When you bite into a hot chili pepper, or a Hungarian red pepper, you first feel a sting and then a burning in your tongue. This is because a substance in the pepper called capsaicin is binding to receptors on pain fibers in the tongue. The capsaicin makes these nerve fibers release their contained neurotransmitter, a short protein or peptide called substance P. So the sting you feel, and then the burning, comes from pain signals transmitted through these very rapid firing nerves. As you eat more hot chili peppers, some of these nerve fibers die off, and so the more you eat, the more you need to eat in order to feel the zing.

Substance P–containing nerve fibers form a rich and lacy lattice-work in skin and joints. Where they are particularly dense, pain is most sensitively felt. During inflammation, the surrounding irritation can cause these nerves to empty their packets of substance P into the inflammatory site—the joint space, for instance, in arthritis. And when this happens, substance P itself makes inflammation worse. Because, it turns out, substance P attracts and then turns on immune cells.

In the 1980s and 1990s, immunologists such as Ed Goetzl, Don Payan, and Jon Levine in California, and Andres Stanisz and Joel Weinstock in Canada, decided to test the effects of substance P on immune cells grown in tissue culture and measure it at sites of inflammation. They did this because when substance P, and peptides like it, are injected into skin, a wheal and flare develops—a hive, just like what people get when they touch or eat the thing to which they are allergic.

These scientists wondered whether substance P could activate macrophages, those garbage-collector cells that gobble dirt. When Payan and Goetzl added substance P to macrophages, they found that the cells gobbled more. The cells had become activated, and when activated, they started spewing out their own transmitter mole-cules, those chemical protein signals called cytokines. They made interleukin-1, and this in turn made lymphocytes make interleukin-2. Others later searched for and found substance P in inflammatory cells and nerves at sites of inflammation. And so, a neuropeptide that in the nervous system communicates pain, at sites of inflammation was calling in and activating immune cells, and amplifying inflamma-tion. This sequence fits with what we know of injury: an inflamed site hurts where it looks inflamed. It goes back to the old Latin description of the signs of inflammation: dolor, rubor, calor, turgor—pain, redness, heat, and swelling. All these signs come from the ef-fects of those nerve chemicals and immune molecules spewed out in inflamed tissue.

At the same time, John Bienenstock, an immunologist, was study-ing mast cells at McMaster University in Hamilton, a small university town in the western part of Ontario. Mast cells are a type of white blood cell that releases a chemical called histamine during an allergic reaction. Anyone who has had an allergic reaction will know what histamine is and how it makes you feel. It is the chemical that causes skin to swell up and become itchy when you get a hive. It is the chemical that makes you sneeze and makes your nose run and eyes water when you inhale pollen or dust. And it is the chemical whose action is blocked by those pills that you can buy in any grocery store for allergy symptoms—antihistamines. It is released from small sacs contained in mast cells, sacs that when stained with the right dye look like blue polka dots inside these cells. When an allergic person comes into contact with an allergen—a protein or chemical that causes allergies—the mast cells' histamine-containing sacs empty. It is the released histamine that triggers the itching, redness, and swelling of the allergic reaction.

Bienenstock wondered whether the itching results from irritation of nerve endings. He began to observe mast cells in tissues, and he noticed that they accumulate around nerve endings of nerves that contain the nerve chemical serotonin. Bienenstock began to do time-lapse photography of these nerve endings. Under the microscope lens of the time-lapse camera, he could watch the mast cells crawl along nerve endings and attach themselves closely to the nerve fibers, like barnacles on a rope. Even under an electron microscope, the connection between mast cells and nerve endings bore a striking resemblance to connections between nerve endings themselves—the synapses. These connections between nerves, called synapses, are the tiny gulfs of fluid between cells across which information, transmitted as electrical charges along nerves, must jump. But the electrical charge can't jump across the space, and so, instead, when the charge reaches the end of the nerve, it empties tiny sacs of chemicals, the neurotransmitters, into the watery space. These neurotransmitters spread across the space to the opposite shore, where another charge is triggered to go racing down the next nerve, until the sequence in the circuit is complete.

Why couldn't these immune cells be communicating with the nerves against which they nuzzled by emptying the contents of their sacs into the fluid space between? In fact, it turns out that not only can chemicals released from nerves activate mast cells, but mast cells also can activate nerves. When mast cells in the intestinal lining empty their contents after exposure to an infectious stimulus, they activate nerve endings lying nearby. This may be one of the ways that nerve-generated cramping, bloating, and diarrhea are perpetuated in irritable or inflammatory bowel disease. So here is yet another way in which the immune system and the nervous system communicate, and yet another way in which the nervous system might regulate disease.

The striking thing about all the effects of such nerve chemicals is that most seem to activate immune cells and so increase inflammation. There is one group of nerve chemicals, however, that, when added directly to immune cells, turn down their activity and could directly shut off inflammation. These are the opiates. Whether drugs or endogenously manufactured proteins, the opiates, morphine, endorphins, and enkephalins all dampen the ability of immune cells to fight infection. The way these compounds exert their effects is by first binding to receptors on the surface of immune cells. So here's another symmetry of nature, another level at which these two great systems communicate: receptors for nerve chemicals are found on immune cells, and nerve chemicals change the way immune cells function.

Since so many different chemicals that come through nerves coursing down the spinal cord from the brain can change the way immune cells work—turn them up or turn them down—it is logical to expect that changing the flow of such nerve chemicals by any mechanism could alter inflammation. So, if interfering with the flow with surgical cuts or drug treatment can block arthritis, why couldn't conditioning and belief do so as well? The missing link to complete this circle is an understanding of what controls the hormones and nerve chemicals of our brain's stress and relaxation pathways. If learning, conditioning, ritual, prayer, and meditation downshift the stress response, decrease stress hormones, and allow enkephalins, endorphins, and other immunosuppressive molecules to play a greater role, then such molecules might also shift the balance of nerve chemicals flowing down the spinal cord from those that induce pain and inflammation to those that tone it down. Although this has not yet been proved, it just remains in this old-new science systematically to define the paths by which thoughts and beliefs formed by learning and association in the cortex of the brain could send signals down through spinal cord and nerves to change immune cell function and disease at sites of inflammation.

CHAPTER · 10 ·

How the Immune System Changes Our Moods

———◀◉▶———

Think once again about how you feel when you are sick. The grogginess you feel, the sleepiness, the not wanting to move, the loss of appetite, the loss of will and strength, the sometimes sadness, and the fever—these are all caused by cytokines, the molecules released from immune cells as they try to fight off the infection. Even if the flu you have is in your stomach, those feelings are generated not in the belly but in the brain. It is your brain that tells you when to eat and when to stop—the signal comes from centers in the hypothalamus, not from the stomach or the mouth. Similarly, it is fever centers in the hypothalamus that set your body temperature, like the furnace thermostat that sets the temperature in your home. Other parts of the brain make you sleep, and motivation, urges to move or not to move, comes not from muscle but from your brain.

If these feelings come from the brain, but the germs that cause infection are in the body, how could the feelings and behaviors happen? In some circumstances, the brain itself is infected with the virus or bacteria (encephalitis), but in most cases when we feel sick, the bacteria and viruses are kept out of brain, wreaking their infectious havoc only in the body's tissues. Some germs infect only the liver, causing hepatitis; others infect just the lungs, causing pneumonia; still others attack the stomach or intestine, causing gastroenteritis. So why is it that in all these infectious illnesses, besides the symptoms we feel that are specific to the infected organ—coughing in pneumonia, diarrhea in intestinal flu—there is also a general feeling of illness? The French word used in clinical medicine to describe this general feeling, "malaise"—meaning literally "ill at ease"—captures its essence. These feelings are so much a part of every illness, and so taken for granted, that until very recently they were dismissed by physicians and scientists as having no importance. In part this stemmed from the fact that they serve little diagnostic value, being so general in nature, other than to indicate to a physician that the individual is indeed sick. A person with fever, muscle aches, and malaise might have a viral or bacterial infection or might have an infection in the liver, lung, or kidney. Or they might have no infection at all, but instead have an inflammatory illness such as rheumatoid arthritis. It is other kinds of symptoms and tests that tell a physician the origin of the illness and, from that, give an inkling of how to treat it.

And yet it is the very generality of these symptoms that holds the clue to their origin and cause. When Hans Selye first recognized the importance of these general symptoms in 1925, he was frustrated at the medical community's rejection of their significance. But when, finally, in his own experiments, he had the insight to recognize that it was the common dirt in his preparations that produced a general syndrome in his animals, he himself was overlooking the real hormones that caused the syndrome. By focusing on the output end of this response—the stress hormones from the pituitary and adrenal glands, which were activated by the dirt—he overlooked the other set of hormones, made by immune cells, that start the whole cascade. These are the cytokines, the interleukin molecules. For contained within Selye's "dirt" must have been bits and pieces of bacteria, the fatty parts of their cell walls. These parts of germs are in all dirt, and they are among the most powerful triggers of immune cells' cytokine production.

But there is a puzzle in this scheme. How do the cytokines, which are floating around in blood or tissues of the body, get to the parts of brain that make you feel sick and trigger the reaction? Neuroscientists were sure that these molecules didn't or couldn't reach such sites because they are too big. The reason it seemed impossible to imagine that big molecules could seep from blood to brain is that between blood vessels and brain tissue there is something called the blood–brain barrier, an impermeable wall that, like the Great Wall of China, surrounds the brain and keeps out large invaders (viruses, bacteria, cells, and even proteins not made there).

If you look closely at blood vessels in the brain, you see a very different pattern of cells than in blood vessels piercing other tissues. Instead of loose-knit cells lining these threadlike cylinders, the cells lining the brain's blood vessels have special footlike parts that lock together and overlap to form an impermeable boundary. Even at the highest magnifications of an electron microscope, where you can see the black fuzz of gluelike molecules in these tissues, there are no gaps, no windowed areas where any sort of large molecule might pass. Under the electron microscope you can measure the size of cells in widths of atoms. Between blood and brain, there are no gaps larger than a few such widths. Molecules bigger than this size couldn't possibly get through. Or so it was thought.

Then pharmacologists began testing the size of molecules that *could* cross from blood to brain. One way to do this is to inject colored dye, or chemicals of different sizes tagged with radioactive atoms, into the blood of rats or mice (the blood–brain barrier works the same way in rats as in humans). If these tagged molecules can be measured in brain tissue, it means they have leaked into the brain. Such experiments showed that a single molecule such as an amino acid could easily get across, and so could short chains; but the longer the chain of amino acids, the less likely it was to cross. When the chemical structure and size of the cytokines were determined to be made of much longer chains of amino acids, it seemed obvious that these molecules, many times too large, could never cross. Yet these large molecules, released into blood and tissues, do affect the brain's behavior. So there must be chinks in the barrier, or other ways for cytokines to signal the brain.

It turns out that there are some chinks—leaky parts that show up right near the hypothalamus. But it seemed unlikely that these small leaky places could account for all the effects of cytokines on the brain. There had to be another way for such molecules to send a signal. One clue to how they do this comes from how we treat a fever.

—◀○▶—

In the 1500s and 1600s, the kings of France sent explorers to the New World—the *coureurs des bois* or "runners of the woods." They came on ships across the Atlantic at first with Jacques Cartier, then with others, to the Grand Banks off the coast of Newfoundland. They traveled up the St. Lawrence River, from its mouth, past the escarpments that became Quebec City, past the rapids at Montreal they thought to be the gateway to China (and named Lachine, French for "China"). They penetrated deeper to the Great Lakes—Ontario and Erie, Huron, Michigan and Superior. With canoes and native guides, they followed the rich network of rivers from these lakes to their sources. They sent couriers down small streams that connected to other river systems, through bodies of water like Vermont's and New York's Lake Champlain. At what became Chicago, they were able to pass through low marshes from the Great Lake system to the Mississippi River valley. In this way, they fanned out over half the continent, portaging ever deeper into woods in drainage basins of the Mississippi, St. Lawrence, and Ohio River valleys. They left their mark in all these places in the French names of towns that still remain—towns as far apart as Portage La Prairie in Manitoba, Canada; Grand Portage in Minnesota; Florrissant ("Flowering Place") and Creve Coeur ("Broken Heart") outside St. Louis, Missouri—a city itself named for a French king, Louis the IX, who had been made a saint.

The prize these trappers sought to bring back to France was beaver pelts. They were as valued in the French kings' eyes as gold. Much as they had sought the silks and spices of the Orient, these kings hungered after the luxurious soft warmth of beaver fur for making rich felts and cloaks and hats. But there was something else that made these animals so prized. It had to do with a secret about the beaver that the native guides had taught the French explorers. Inside the neck of the beaver, there is a small gland that, when carefully excised, dried, crushed, and taken as a powder, can cure a fever and rid the sick of headache and of pain. What magic substance was this and where did it come from? It came from willow bark. When beavers chew on willow to bring down young saplings to build their dams, the chemical constituents of the bark are concentrated in this special gland inside their neck. One of these chemicals is called salicylic acid. We know its derivatives today as aspirin.

Aspirinlike compounds were used for centuries to cure fever and to take away the symptoms of infection. They were used long before

German chemists in the Rhur River valley, at a company called Bayer, synthesized them in 1899 and called them aspirin. And aspirin in turn was used for decades before the British scientist John Vane discovered that aspirin blocks the enzyme that makes something called prostaglandins, which can cause a fever. Vane found that aspirin blocks the body's chemical machinery that breaks down long chains of fatty acids to make the small molecules of prostaglandins. (Vane's discovery not only paved the way for the design and discovery of new prostaglandin-blocking drugs with the same effects; it won him the Nobel prize in 1982.) Later, Charles Dinarello found that molecules such as interleukin-1, when injected into brain or blood, also cause fever. And aspirinlike drugs were used before it was discovered that when interleukin-1 binds to its receptors on the cells lining blood vessels in the brain, it triggers the internal machinery of such cells to turn on prostaglandin-producing enzymes and pour out prostaglandins into the surrounding brain tissue.

One important way that cytokines such as interleukin-1 can signal the brain, then, is by turning on a blood vessel's machinery to make prostaglandins. These molecules, small enough to pass through the blood–brain barrier, can diffuse out into brain and produce their effects. It also turns out that when cytokines bind to blood vessel wall receptors, they trigger other enzymes to make other tiny molecules— gases that can leak out quickly into brain tissue and also act as "second messengers" in brain. This other enzyme is called nitric oxide synthase. The gas that is produced when this enzyme breaks down nitrogen-containing amino acids is nitric oxide. (The three scientists who discovered the phenomenon that a gas can act as a signal molecule by diffusing through tissues were awarded the Nobel prize in medicine for their discovery in 1998.) Endocrinologists Samuel "Don" McCann and Valeria Rettori later found that this gas can also turn on brain cells to make hormones, including the stress hormone CRH.

In a kind of relay race from large to small, the message that there is infection in the body spreads deeper into brain tissue. Yet not all the effects of cytokines on the brain are blocked by prostaglandin-blocking drugs, and therefore it would seem that another mechanism, beyond prostaglandins, is capable of transmitting these signals. Moreover, some effects of cytokines on the brain occur much too quickly to be explained by cells first making a small second molecule that then diffuses outward. There had to be other ways for immune molecules to produce their effects on the brain.

—◄○►—

How can an immune molecule, released during an infection in the belly, signal the brain in less than seconds? How could the signal travel from the center of the body to the head so fast? In the early 1990s, two groups of scientists, one in Boulder, Colorado, and the other in Bordeaux, France, asked this question and found that the immune signal is carried along the fastest route that exists in the nervous system: the nerves themselves. One group was studying pain pathways in the spinal cord, while the other was studying sickness behavior. Both published papers within a few months of each other showing that immune signals from the belly travel to the brain along the vagus nerve.

Until these two groups realized that it also carries immune signals, the vagus nerve was known to physicians and physiologists for its importance in digestion, in regulating heart rate, and in regulating breathing. It is one of the main structures of the parasympathetic nervous system—the part of the nervous system that regulates the functioning of the viscera. Signals travel in both directions along the vagus nerve—from gut to brain and brain to gut. When we feel something viscerally what we are describing are the bodily sensations that come through the vagus nerve.

This large nerve trunk sprouts from organs down in the belly: the liver, the intestines, and the fatty, lymph node–laden apron that surrounds the gut, the peritoneum. As it winds up through the chest, it takes on other branches from the lungs and heart and the great blood vessels in the neck. The vagus then enters the brainstem, where it ends in a kind of switching station, named by the anatomists the nucleus of the tractus solitarius—the solitary nucleus—because it is all alone, away from the other grayish shapes in the brainstem. When it is stimulated by small bursts of electricity, electrical signals travel very quickly, in thousandths of a second, up along the vagus to end on collections of nerve-cell bodies in this solitary switching station. In fact, when the vagus nerve is stimulated, this nucleus up in the brainstem is the first place that lights up. It comes to life with rapid-fire bursts of electrical impulses and then with the quick production of messenger RNA—the molecule that tells you proteins are being made from genes. From there, the next relay set of nerves is activated and more electrical signals pass to other parts of the brain.

The vagus nerve can be activated in many ways—with pressure, by electrical charge, by drugs, and by physiological processes, such as eating, defecating, or motion sickness. Activation of the vagus nerve can make you faint—by switching on parts of the brain that lower blood pressure and slow your heart rate. This is called a vaso-vagal attack. Sudden pressure on the vagus nerve, where it winds around the great artery in the neck, the carotid artery, or where it originates in the belly, can also make you faint. Other activation of the vagus can make you vomit. This is why if you are punched in the abdomen, you feel faint and nauseous. The vagus nerve has a lot to do with the gut—with the stomach, with digestion, with the liver. It is the great nerve cable that mediates those sensations that we associate with "gut reactions." Its role in digestion was one clue that pointed to the vagus nerve as a route by which cytokines in the belly could signal the brain.

In Boulder, Colorado, two neuroimmunologists, Steve Maier and his wife, Linda Watkins, were studying the effects of cytokines on pain. Inflammation and the cytokines produced by inflammation quickly cause pain and fever, even if the inflammation is only in the belly. Pain signals generated by other triggers are carried through the spinal cord. To test whether the spinal cord also carries cytokines' pain signals to the brain, Maier and Watkins injected interleukin-1 into rats' abdomens, and then measured electrical signals in the spinal cord and in parts of the brain just above the cord. They found that these signals increased after the injection. But how did the cytokine signals get from the belly to the spinal cord? Could it be that it was the vagus nerve that carried these signals? To answer this question, Maier and Watkins cut the vagus nerve just under the rats' muscle that separates the chest from the abdomen, the diaphragm. The electrical signals didn't get through.

But how and why would an immune molecule such as interleukin-1 activate the vagus nerve under the diaphragm? It didn't make sense that under day-to-day circumstances the body would need a mechanism to sense immune signals in this place. One evening, after a long session of talks at a scientific conference, Maier and Watkins sat discussing this question at the hotel's bar. From a nearby table, they heard simply, "It's the liver." It was the voice of Dwight Nance, a Canadian friend and colleague and expert in the anatomy and physiology of digestion, who had been listening to their conversation. "Why the liver?" they asked. One reason, Nance pointed out, is that the liver is richly innervated by the vagus nerve. The other is that the liver is filled with macrophagelike cells that make cytokines. So the branches of the vagus nerve that course through the liver would be a logical route along which cytokine signals could be carried.

As soon as Maier and Watkins returned to the lab, they did another experiment in which they cut the vagus nerve, just above where it sprouts from the liver. They found that cutting it there blocked the pain and electrical signals in the brainstem brought on by inflammation. So they concluded that at least some of the effects of interleukin-1 can reach the brain by rapid electrical signaling along the vagus nerve, especially along the branches that come from the liver. But how did the interleukin-1 activate this nerve not when it was made in the liver but when it was injected into the belly? There had to be some place where it could bind. Maier and Watkins then found small nodes of tissue stuck close up on the branches of the vagus, named long before by the anatomists who discovered them for their location next to the spinal column—the paraspinal ganglia. Others had shown that these ganglia contained receptors for neurotransmitters that the vagus used. Why shouldn't they also contain interleukin-1 receptors? If they did, it could explain where interleukin-1 was binding to produce its electrical perturbation of the vagus nerve. So, Maier and Watkins stained these paraganglia with antibodies for interleukin-1 receptor. They found what they were looking for.

Meanwhile, Robert Dantzer, in Bordeaux, France, was studying the effects of cytokines on rats' behavior. After injecting interleukin-1 into the rats' bellies, he observed decreased movements, increased sleep, and loss of appetite—all symptoms of sickness behavior. Although Dantzer's research interests were in behavior, he had started out as a veterinarian, with many hours of surgical and anatomical studies in his training. From this, he knew that the vagus nerve was important in transmitting all kinds of signals from the organs in the belly. Around this time he had heard about Watkins and Maier's experiments on pain. Could the vagus be involved in sickness behaviors, too? To test this hypothesis, his colleague Rose-Marie Bluthé performed surgery to cut the vagus nerve and then injected inter-leukin-1. The rats acted as though they had not been exposed to any-thing at all—they showed no sickness behaviors and had no fever. Bluthé found that this was true no matter where she cut the vagus nerve, whether at its small twiglike branches sprouting from the liver or on its main abdominal trunk. So it seemed that at least some of the effects of cytokines could be carried to the brain, not by cyto-kines themselves crossing into the brain, not by second messengers such as the prostaglandins, but by electrical impulses generated in the belly and passed at lightening speed up along the vagus nerve.

Clearly another route by which cytokines in the body could sig-nal the brain had been worked out. But this still did not account for all the ways that cytokines could signal the brain. It turned out that there was yet another way.

◄o►

If you want to test whether a membrane is permeable, you can do an experiment in your own kitchen to find out. All you need to do is fill a bottle with colored water and cover it tightly with the membrane—tight enough that you can be sure there is no leakage around the edges. You then immerse the bottle in another container filled with clear water and watch to see if the color passes from one bottle to the other. If the color from the inner bottle seeps through the membrane to the outer container, you know the membrane is porous. If the two liquids keep their separate colors, the membrane is impermeable.

This is effectively what Bill Banks, a neurobiologist at Tulane University, did to find out if interleukins crossed the blood–brain barrier. Instead of using colored water, Banks used radio-labeled interleukins, which he injected into the blood of rats; he then meas-ured how much radioactivity seeped into the brain. Some did get through, though very little. At first, other scientists thought that this could just be coincidence. They argued that perhaps the radioactive label was coming off the protein, and so diffusing unattached. Or maybe the blood vessels were leaky, as if in the kitchen experiment you hadn't tied the cellophane on tight enough. There are parts of the blood–brain barrier that are leaky, those conveniently located near parts of the brain that interleukins affect—especially near the hypothalamus. So maybe interleukins could get through these areas, windowed with larger holes than all the rest.

But Banks was convinced that interleukins were not just leaking through but might be actively carried across the barrier, as some other molecules are, by carrier proteins sitting in the cell membrane. Such proteins had been found in other cells—in heart muscle, for instance, where they sit in membranes and work like waterwheels, catching passing sodium or potassium ions and carrying them inside the cell. Persisting in his experiments to test this theory, he eventually found that human IL-1 injected into blood could be detected in mouse brain tissue at a fast enough pace to show that it had to have been pumped into the brain and was not just passively seeping through.

—◄○►—

Today it is clear that there are many routes by which immune molecules made in the blood can get to the brain and stimulate brain functions. But besides the fever and malaise that we feel when we are sick, can these molecules otherwise directly affect our moods? Can they explain the sadness that sometimes accompanies illness? This is a hard question to test in animals, because we can only observe their behavior, we can't ask them how they feel. And the question is only just beginning to be tested in human subjects. What animal studies do tell us is that if you inject interleukin-1 into the belly of a rat, and then put it into a cage with other rats—back into its social environment—there is a change in its behavior that cannot be accounted for just by fever or sleepiness. It no longer interacts with other rats; it is neither aggressive nor affiliative. In fact, it acts the way we often feel when we are ill—withdrawn from all the world. So it may well be that cytokines also affect the parts of the brain that bring on these behaviors, and they might even affect the parts of brain that change our mood to sadness. This would make sense because the parts of the brain that are most directly connected to the immune system, on both the receiving and transmitting ends, are those parts that regulate the emotions: the stress center in the hypothalamus, and other mood centers, such as the amygdala's fear center. In fact, neuropharmacologist Adrian Dunn at Louisiana State University in Shreveport found early on that many neurochemicals we know to be involved in mood shifts—chemicals such as serotonin, dopamine, and norepinephrine—all change in different brain regions in rats exposed to interleukin-1. And Jay Weiss, a neuroendocrinologist in Atlanta, Georgia, found that after interleukin-1 is injected into the belly of rats, fresh interleukin-1 is also made in parts of the brain that regulate depression. Here is yet another way that cytokines could get into the brain to affect our moods: besides arriving by signals sent from the body, they can also be made by cells inside the brain.

–◀○▶–

There are so many ways that cytokines from the body can signal the brain: passing through leaky chinks in the blood–brain barrier; being actively pumped in; triggering small signal molecules that then diffuse into brain tissue; signaling through the vagus nerve. Yet while all these routes show that there are many of lines of communication, there is something striking about this elaborate system. It seems, in Shakespeare's words, that perhaps the lady doth protest too much. In fact, this elaborate system, a series of checks and balances, would almost appear to have been set up to keep these molecules out, rather than to let them in. When you think about it, there are an extraordinary number of roadblocks built into the system to prevent or at least control the flow of cytokines into the brain. Why might this be? Could it be that when such immune molecules get into the brain tissue in too large a quantity they could be harmful, actually damaging nerve cells? What might happen if an immune cell, spewing out cytokines, accidentally, or in disease, squeezed its way into the brain? And what about such molecules made right in the brain? Do they help or do they harm?

In the immune system, cytokines generally stimulate other immune cells to grow, divide, or mature. And it turns out that they have the same effect on young nerve cells. So, in a developing nervous system, these molecules are, as they are in the immune system, growth factors, important in maintaining cell growth and survival. But it also turns out that, in other circumstances, these molecules can do the opposite to nerve cells. If mature, no-longer-developing nerve cells are bathed in a soup of interleukin-1, the cells will die at a more rapid pace than usual. The natural process is that once nerve cells growing in the tissue culture dish have multiplied to fill the dish—in dense sheets of cells touching other cells—the oldest cells begin to die; and they die at a regular pace, so that the dish never becomes too crowded. If you add to these cultured nerve cells an antibody or a drug that blocks interleukin-1, you prevent that natural nerve-cell death. In health, there is a balance between nerve cell death and cell growth, so that neither happens too much or too quickly.

Nerve cells can die for many reasons. They can shrivel up and die from starvation, from too few nutrients; they can burst apart if they imbibe too much fresh water without the necessary dose of salt; and those long axon feet, which connect nerve cells to other nerve cells, can slowly die off if the filamentlike railway system that transports nutrients from cell core to tip is poisoned. But there is yet another way that all cells, including nerve cells, die. It is called programmed cell death—cell suicide, really—and it is called this because it is programmed into the genes. Once the genes that trigger cell death are activated, a sequence of events is set into motion in which enzymes within the nucleus begin systematically to clip DNA into small pieces. This active cell death is called apoptosis, and it turns out that the way cytokines make nerve cells die is by triggering apoptosis.

So, what happens if an active immune cell, like a macrophage, spewing out its cytokines, gets into the brain during disease? Do nerve cells, caught in the crossfire, die? This does, in fact, happen, and one disease in which it happens is AIDS. Another is Alzheimer's. Some patients with AIDS can begin to lose their memory and can go on to develop an illness that in many ways resembles Alzheimer's disease. In these and other neurodegenerative diseases there is a standard pattern and progression of memory loss. At first just a few words and events are forgotten here and there—so few that the person cannot tell if the forgetting comes merely from fatigue or stress or distraction. Indeed, these situations in normal individuals often involve some holes in memory. But in an Alzheimer patient, the memory loss relentlessly worsens. Eventually these patients can no longer remember how to perform the simplest tasks of daily living—how to open a can of tuna or peel the wrapper from a wedge of cheese. This is called dementia. Then they reach the stage at which they can no longer recognize loved ones, can't feed themselves or swallow. And finally they succumb.

Pathologists looking through the microscope at the brains of deceased AIDS patients with such dementias can discern collections of giant cells with many nuclei—sometimes more than a dozen nuclei. These cells have foamy cytoplasm surrounding the clusters of nuclei that in many ways resembles the cell soup of macrophages in the tissues or monocytes in the blood. The reason these cells look alike is that they are related; they are, in fact, more than close cousins. These giant, multinucleated cells in the brain are formed from many macrophages, all infected with the AIDS virus, that have fused together into one giant cell. These cells find their way into brain tissue, creeping through chinks in the blood vessel walls. Just as do macrophages in other body tissues, these huge cells are making cytokines, pouring them out into their surroundings. If a pathologist were to bathe the microscope slide with black-ink-tagged antibodies that binds to interleukin-1, the interleukin molecules would suddenly become visible in the tissue surrounding these cells. The image that the microscopist would see is a concentration of black spots close to and on top of those giant cells—and nowhere else around. It is likely that such concentrations of cytokines, around invading macrophages, trigger the nerve-cell death that accounts for the symptoms of dementia in these patients.

What about in Alzheimer's disease itself? Do cytokines explain those memory-loss symptoms as well? Under the microscope, the brains of patients who died from Alzheimer's disease are riddled with holes. These are not actual holes, as in Swiss cheese, but spots where the neat and ordered nerve pathways and connections that should be there suddenly lose their way and degenerate into filamentous, hair-ball-like tangles. And, yes, within these plaques, as they are called, there are accumulations of immune cells, and with the immune cells, accumulations of cytokines.

However, just because there are excess cytokines in the brain in these dementing diseases, AIDS and Alzheimer's, does not mean that these molecules are necessarily the culprits killing off the nerve cells. Other kinds of experiments do suggest that this is so. One way to prove this connection is to add cytokines to nerve cells in a tissue culture dish, and then simply count the number of dead nerve cells. It turns out that very low concentrations of interleukin-1 can kill nerve cells in tissue culture. Another way to prove this is to genetically engineer mice that make excess cytokines in their brains. Ian Campbell at the Scripps Research Institute in La Jolla, California, performed such an experiment and found that such genetically engineered mice can develop illnesses of nerve degeneration—loss of ability to learn, loss of motor function and of balance. Still another way to prove that such cytokine molecules can cause nerve-cell death is to block their effects with drugs and see whether cell death is prevented. Nancy Rothwell in Manchester, England, did this in animals with stroke, the death of nerve cells that results from a sudden blockage in blood supply to the brain. She found that the naturally occurring molecule that blocks interleukin-1 from binding to its receptor, called the interleukin-1 receptor antagonist or IL-1ra, can decrease the size of a stroke in lab rats by more than half, when given early enough in the course of the stroke. And drugs such as aspirin, which block the effects of cytokines by blocking formation of those small second-signal molecules, prostaglandins, also reduce the risk of Alzheimer's dementia. Putting all these studies together, it seems that cytokines do cause nerve cells to die in certain circumstances, and treatment with drugs that oppose the effects of cytokines can help to prevent such cell death.

While much of the mechanism of cytokines' effects on the brain has been pieced together, there are still many questions to be answered. For example, do the cytokines that show up in the brain during these illnesses always come from invading immune cells? We think not. The reason is that brain tissue is made up not just of nerve cells but consists also of a scaffolding of many other kinds of cells onto which nerve cells are attached; and if these scaffolding cells are grown in tissue culture, cytokines can be measured in the soup that bathes them. In other words, these cells actually spew out cytokines—many different cytokines and in large amounts. If this is happening in normal conditions inside the brain, then cytokines can't be all bad—they can't be just killing nerve cells. Indeed, as we saw earlier, cytokines prevent the naturally occurring cell death in immature nerve cells grown in tissue culture, which suggests that they must in some circumstances be needed, especially in a developing brain. In fact, in the adult nervous system that has been damaged by stroke or trauma—such as spinal cord injury—nerve cells regenerate just as they do early in life. And in these circumstances, cytokines and the immune cells that stream to the site of the break also guide regrowth of nerve cells. Michal Schwartz, an Israeli neuroscientist at the Weizmann Institute, used this discovery to reestablish sight where optic nerves had been cut and to reestablish locomotion in rats whose spinal cords had been damaged. In the constantly reshaping, growing, changing nervous system, cytokines play a vital role in both preventing and causing nerve-cell death. It is as if these molecules are sculptors that help to chip and chisel, or add small thumbnail bits of nerve tissue here and there. And if the process goes out of balance—too much chipping or too many bits added to disrupt the pattern—the final intricate relations between nerve cells and their many connecting paths may be disrupted and lead to disease. But they may be also be crucial in reestablishing connections and function.

What does all this tell us? Most importantly, that diseases we thought previously unconnected seem to be linked by a common thread. Those illnesses characterized by dementia and nerve-cell death are connected because the final bullets that kill these cells are cytokines. Knowing this will help in the development of new treatments for such illnesses, the development of drugs aimed at blocking cytokines from killing nerve cells before their time. As more is learned about the other side of cytokines' effects—the enhanced survival of growing nerve cells—new treatments can then be developed to help heal damaged nervous tissue and for use in new and yet undiscovered illnesses in which this balance is perturbed.

But cytokines are not expressed in the brain only at times of growth and of repair. Neuroanatomist Clifford Saper at Harvard University School of Medicine first showed spidery nerve pathways of interleukin-1 in the human brain, and neuroanatomists such as Paul Sawchenko at the Salk Institute in San Diego mapped out the distribution of receptors for this cytokine in the brains of rats. The presence in the brain of these intricate pathways of both the immune molecule and its receptor suggests that these molecules may be playing a role not only in disease but in normal brain functioning as well.

Taking a cue from those more ephemeral effects of cytokines on mood and their effects in triggering feelings of sadness when we are ill, we not only can begin to understand the molecular underpinnings of sickness behavior but can also determine whether cytokines expressed within the brain might also trigger feelings of sadness in the absence of infection. For it could be that, in some cases, cytokines out of balance in the brain could lead to illnesses such as depression. If that is so, whole new lines of therapy could be developed to treat depressions in which these molecules might play a role. It is these sorts of ideas, previously inconceivable, which this science is now poised to test.

CHAPTER · 11 ·

Prometheus Unbound

What Does the Future Hold?

————◄o►————

*T*he edges of the marble steps are sharp and square, not worn thin by the footsteps of the ages. At the top of the steps is a tall brass door in what seems to be the entrance to an Egyptian tomb: a low, two-story, square limestone building, decorated along the roof and on the outside wall with weathered copper gargoyle trim and friezes. Inside the dark interior are metal gates carved with the twelve signs of the zodiac and tessellated winged sphinxes giving onto a low-ceilinged antechamber, like a vault, dimly lit by tall torch lamps. The wooden beamed ceiling is painted in surprisingly bright tones. Through the next metal-gated arch footsteps echo on the polished green stone floor of the great hall, a square court surrounded by a colonnade of green serpentine marble Corinthian columns. The eye is immediately drawn to the center of this court. There, suspended by an almost invisible wire, a brass ball hangs from the domed ceiling two stories above. Almost imperceptibly, the sphere moves, tracing an invisible path over the pedestal below. In perpetual motion, slow and sure, this pendulum swings with the turning of the earth and seems, with the passage of the day, to change its direction when in fact it is the earth that is shifting around it.

In a way, this amazing device is symbolic of the ways of science—always moving, yet charting a course that is determined by the world

around it, and, in the end, never really moving away from the principles of nature that guide it. New ideas and discoveries, new concepts
that turn a previous generation's thinking on its end, are never easily
accepted. They require much swinging of the pendulum back and
forth, until some new equilibrium is established. So, it is fitting that
this symbol hangs from the pantheon-like dome, not of a Roman or
Greek temple, but of the entrance hall of the National Academy of
Sciences in Washington, D.C.

This modern temple of science, built in the early 1920s, incorpo-
rates in its architecture bits and pieces of many of the great institutions of Western science throughout the centuries. Besides the echoes
of Egypt, Greece, and Rome, there is a steep row of seats lining the
mezzanine above the court—looking for all the world like Padua's
anatomical dissecting theater. The domed ceiling is painted with a
faux-mosaic solar system—sun in the center, concentric circles of
planets and symbols of the sciences around it—in a Renaissance style
that could have adorned a cathedral dome.

We have moved on from the unquestioning awe of science that
prevailed when this building was designed to a different soil: one
where the public demands more justification and relevance of the
work, one where the clamor of the popular culture pushes a sometimes reluctant scientific world to turn its esoteric eyes on the people's more immediate and practical concerns. The mind–body connection, first a child of the popular culture, is finally making its way
into these echoing halls as more and more scientific evidence accumulates to strengthen the proof of these connections. And even as
the proof in humans is still incomplete, the urgency to apply these
principles to preventive medicine—to improve our health—is at least
spawning questions to be asked by those public institutions that must
decide now, not several decades hence, what to do.

—◄o►—

To try to predict the future in science is always a risky business. To
contemporaries, fantastic predictions seem impossible, and to later
readers the context of the thoughts, rooted in assumptions of the day,
seem dated. Sir Francis Bacon, a barrister and politician, and lord
chancellor of England in 1618, was himself only a dabbling scientist.
But Bacon saw the great potential of marrying the goals of the state

with the needs of science. He lived in an era where new discoveries of new lands were fresh in peoples' memories. And he recognized the power of applying scientific principles to navigation, health, and other public concerns. In his global way of thinking, he envisioned a time when the scientific approach to observation and discovery would be applied for the universal good and when the research of scientists would be supported by the state. And yet we have to smile at his descriptions of this idyllic utopia in his essay "The New Atlantis," so shaped are they by his post-Elizabethan times. The imaginary island he described was home to a wise people, saved after the Great Flood, from a previous, more scientifically advanced world that had been mostly destroyed. These people blended scientific principles and social conscience, and they developed institutions that benefited all.

Bacon describes his imaginary ship and the sailors' first encounter, after being marooned at sea, with the Atlanteans. A richly dressed emissary, fluent in ancient Hebrew, Greek, and Latin, as well as Spanish, meets the ship and leads the sailors onto land. Here, the healthy are given well-appointed rooms in a great house and the sick are given special quarters:

> He led us to a long gallery, like a dorture [dormitory], where he
> showed us all along the one side (for the other side was but wall
> and window), seventeen cells, very neat ones, having partitions
> of cedar wood. Which gallery and cells, being in all forty, were
> instituted as an infirmary for sick persons. And he told us withal,
> that as any of our sick waxed well, he might be removed from his
> cell to a chamber. . . .
>
> There were brought to us great store of those scarlet oranges,
> for our sick; which (they said) were an assured remedy for sickness
> taken at sea. There was given us also a box of small grey, or whitish
> pills, which they wished our sick should take, one of the pills, every
> night before sleep; which (they said) would hasten their recovery.

After all were quarantined for three days, and the sick were healed, they were treated to a long discourse by the governor of the island, describing its idyllic and advanced way of life. These people had built deep caves, "above three miles deep," for preservation, refrigeration, and mining. They had tall towers, some half a mile in height, for "the view of diverse meteors as winds, rain, snow and hail; and some of the fiery meteors also. And upon them in some places are dwellings of hermits, whom we visit sometimes and instruct what to observe." There were also artificial wells and fountains for infusions and for prolongation of life. And for the preservation of health:

> We also have certain chambers, which we call Chambers of Health, where we qualify the air as we think good and proper for the cure of divers diseases. . . . We have also fair and large baths, of several mixtures, for the cure of diseases, and the restoring of man's body from arefaction [drying up]: and others for the confirming of it in strength of sinews, vital parts, and the very juice and substance of the body.
>
> We have dispensatories, or shops of medicines. . . . Wherein . . . we have such variety of plants more than you have in Europe. . . . And for their preparations, we have all manner of exquisite distillations and separations . . . through divers strainers . . . ; but also exact forms [formulas] of composition, whereby they incorporate almost, as they were natural simples.

Bacon envisioned a technology-based society in which all endeavors grew from scientific investigation and principle, from mining and weather prediction, to agriculture and animal husbandry, to production of drugs and improvement of health. In Bacon's vision, it was the benevolent leader of this island who oversaw all these endeavors, from his throne room—so much was Bacon still a man of his century.

Bacon was the first statesman to put into action this broad and encompassing view of science and politics, in which discoveries in a technology-driven society would be used for the betterment of all. Just as the anatomists of his day forever put their stamp on physicians' ways of thinking about disease, Bacon changed our assumptions about the relationship of science and society. There is now, in our modern scientific structure, an expectation that the state must underwrite scientific research. And although a parallel universe of science exists in industry, which drives its advances, the funding structure of research in universities still depends mostly on government sources. In the United States funds for basic research come from institutions such as the National Research Council of the National Academy of Sciences, and for medical research, from the National

Institutes of Health. In a utopia, such funds for scientific research should encompass not only the most basic research but also more applied science that can be integrated for the public good.

—◀○▶—

If you pass the pendulum in the center hall of the National Academy of Sciences and continue on, you walk under a mural on the northern wall, a painting of Athena following Prometheus, who is stealing divine fire from the chariot of Helios, the Greek god of the sun. Inscribed on the wall below is a quotation from the fifth-century B.C. Greek play by Aeschylus, *Prometheus Bound.* The quote sings the praises of the sciences, as the source of benefits to the world:

> Harken to the miseries that beset mankind. They were witless erst
> and I made them to have sense and be endowed with reason. Though
> they had eyes to see they saw in vain. They had ears but heard not,
> but like to shapes in dreams. Throughout their length of days without
> purpose they wrought all things in confusion . . . until such time as
> I taught them to discern the risings of the stars and their settings.
> Aye, and numbers, too, Chiefest of Sciences I invented for them, and
> combining of letters, creative mother of the muses' arts. . . . If ever
> man fell ill there was no defence, but for lack of medicine they wasted
> away, until I showed them how to mix soothing remedies wherewith
> they now ward off all their disorders. . . . Hear the sum of the whole
> matter—every art possessed by man comes from Prometheus.

The architects and men of science who placed this quote, carved in stone, so prominently in the center hall of the Academy, espoused a view of science accepted in their time. This somewhat grandiose and egocentric view, which had its roots in Bacon and the philosophy of Descartes, has probably contributed to the public's disillusionment with science and rejection of its ability to heal all society's ills. Yet scientific discoveries have changed the topography of our modern world so much that we can now look back at Bacon's vision with amusement at its simplicity. So why reject it all in cynicism? Why not embrace both sides—the popular need for emotion, and the scientists' need for hard facts and proof?

Through Prometheus's muraled arch you pass into the final hall, a spare modern lecture theater, with rows of red-cushioned seats under a cream-colored geodesic dome. The focus of this hall is not a mural but a stage and a projection screen, where invited scientific speakers present their data at the podium. At public workshops devoted to addressing specific questions, expert panels present their evidence on all sides. The questions posed are usually of practical importance for developing public policy. The premise of the institution when it was conceived was that such policy should be based on as much unbiased knowledge as was possible to gather—a kind of scientific equivalent of Thomas Jefferson's philosophy of political knowledge that spawned the Library of Congress.

And so now, pushed by popular demand, somewhat reluctantly perhaps, the scientific medical world must answer questions that the public asks. Is stress really bad for your health and immune responses? If so, how can we design less stressful work environments and develop ways of coping with such stresses as military service or routine areas of our lives? Does stress contribute to some of the diseases seen in troops returning from a war? How much of the effects of stress are learned and how much is in our genes? Is it too soon to design studies to answer these questions—do we have the tools to answer them now or must we wait?

Technologies are continuously evolving—like your next computer, they are continuously redesigned and upgraded. For physicians to wait to apply the available technologies to practical health questions until all possible technologies are perfected would be like waiting for

---

Left: *Prometheus's grandiose sentiment that he, who granted to man the knowledge of fire, alone possessed the ability to heal ills has long reflected a similar view in the medical community. But this view, which had its roots in the philosophy of Bacon and Descartes, has also contributed to the public's disillusionment with medical science. And so, if the new field of brain-immune communication accomplishes anything, certainly it will be this: to help medical specialists think of the person as a whole—one single body with one single soul; to help physicians speak the language of their patients and listen to them. Through that listening and learning process, we will have crossed the hardest divide of all—the one between the popular culture and science.*

the perfect computer before buying one. You would remain without one. We may never have all the tools to answer all aspects of these questions, but by combining what we have, we can begin.

In Bacon's time, the explosive increase in exploration of the globe, along with new discoveries in all the sciences, led him to envision a more integrated world of science and technology for the greater good. It seems clear that we are positioned at a similar point in a different scientific age, one in which discovery coming upon discovery has led to deeper rather than broader knowledge in each discipline. In the biological sciences, it has led us into the fine workings of the cell, our chromosomes and genes. This dominance of focused thinking, this looking inward of each discipline, has led to some of the greatest discoveries of the scientific era from which we are now emerging.

But if we are to make the leap into the next era of science, we must also include a look outward from each discipline and a reintegration of them all. For now that the threads have been teased apart and examined individually, it is time to reweave the tapestry and discover how it is woven together. To do this, many gaps must be bridged and many new languages learned. As C. P. Snow described in his essay "The Two Cultures," we must drop the arrogance and walling off that comes with increasing specialization—an arrogance born of fear of ignorance of the other's field. And we must learn each other's languages and try to understand how the technologies and approaches to solving problems of one discipline can be applied in order to solve the problems of another.

The science of brain–immune communications is by its very nature a field that does this. It looks inward to the most detailed level of body chemistry and at the same time it looks outward to the larger concerns of health and emotion. It applies technologies that analyze molecules and genes with technologies that image the functioning of whole organs like the brain. It bridges specialized disciplines of basic science such as immunology and neurobiology, and it bridges specialized fields of medicine such as psychiatry and rheumatology. It bridges the basic sciences with clinical medicine and both of these with the intangible but essential input of feeling and emotion. The end result is to make the body and mind whole again.

It was probably necessary to go through the exercise of increasing specialization from the time of Descartes and Bacon until the middle of this century in order to achieve the level of detailed understanding that we have today. But so overwhelmed with detail is each scientific discipline that, in the domain of health at least, sometimes the whole has been lost in the focus on the parts. Perhaps in response to scientists' preoccupation with what seem to be minutiae, disillusioned with the medical community's enthusiasm for shiny technological toys at the expense of human interaction, the popular culture has turned away to seek more seemingly controllable, less alienating ways to health. Herbs and Zen meditation, acupuncture, spas and crystals— "alternative" forms of healing—are sought, paid for by the public with out-of-pocket billions of dollars, approaching or even exceeding traditional health care costs. How much of any benefit from these cures is placebo, how much are the desperate hopes of vulnerable people taken advantage of by clever salespeople? How much is real?

Perhaps the greatest future contribution of this new-old science will be to bring these two poles together once again. In fact, this field is not so much a bridge as a meeting point between science and the popular culture. Hopefully, the skeptical scientist—who in the tradition of Bacon and Descartes believes that unless an effect can be observed and documented, measured and understood, it isn't real— may be united with the person who, when feeling ill or in pain, knows in their guts and heart that what is being felt is real, even in the absence of visible evidence.

The way this emerging science can accomplish such a difficult reunion is by convincing both sides of the value of the other's view. For scientists, the totality of experimental observation, brought together from all sorts of studies—in genes and cells, animals and humans— can lead to a finer and more detailed definition of the intricate workings of connections between the nervous and immune systems. Ultimately through such knowledge, the conundrums of the mechanisms of illness will be uncovered, thus opening up new avenues for previously unthought-of treatments for disease.

This process has already begun, especially in areas of this new science that have been most explored. Thus, knowing that immune molecules can kill neurons has led to a search for such molecules in the brains of dementia victims—people whose memory slowly disappears as in Alzheimer's disease and AIDS, as well as in sudden injury such as trauma or stroke. Finding these immune molecules at sites of nerve degeneration has led to the testing of drugs that block cytokines, such as the interleukin-1 receptor antagonist, to prevent the nerve cell death that occurs in stroke. And since this treatment works, in animals, at least, it proves to the skeptical scientist, in yet another way, that such connections are real and do play a role in healing and in disease. To the patient or caregiver of such patients, it provides a new source of hope that some of their pain and suffering may soon be cured.

The principles underpinning this new science also provide a basis for physicians and health professionals to step back and hear their patients, to recognize that emotions do play a very important role in health and in disease. And this in turn may help physicians to spend more time listening to what the patient has to say. It may help them think of the patient as a whole and treat with words and with compassion before so quickly pushing limb and head and belly through computerized diagnostic tools. And if really listened to and heard, the public may be more willing to accept these otherwise alienating technologies that do help diagnose and cure disease.

At the same time, we can use the technology-driven science of this field to help us begin to think of illnesses, previously thought to be unrelated, as connected, at least in some aspects, by a common thread. Inflammatory illnesses such as rheumatoid arthritis, allergic diseases such as asthma and dermatitis, fatigue conditions such as chronic fatigue syndrome and fibromyalgia, mood disorders such as depression and seasonal affective disorder all show some perturbation of the hormonal stress response. And so, this understanding could lead to development of new treatments for these illnesses. Drugs now being developed to turn down CRH in forms of depression with a switched-on stress hormone response might also be used in joint diseases such as arthritis, where too much local CRH in joints makes inflammation worse. If drugs can be developed to turn down CRH, maybe drugs that turn it up can be applied to patients with inflammatory or allergic illnesses whose stress response is tuned too low. Such drugs might also work for atypical depression—the form of depression with a sluggish stress response.

In addition, new uses for old treatments can be thought of that could not have been conceived before. It turns out, for example, that deprenyl, a drug that acts by blocking enzymes that make adrenaline-like nerve chemicals and is used in psychiatry to treat depression, also makes adrenaline-like nerve fibers sprout and grow in the aged spleen. During the natural process of aging, these adrenaline-like, or sympathetic, nerve fibers die off in the spleen. So, too, do the functions of lymphocytes surrounding them. Treating rats with deprenyl not only makes these nerve fibers sprout to life again but also restores the weakened immune response that comes with age. Another drug, used in endocrinology to block the actions of the milk let-down hormone prolactin, can now be tested against the autoimmune disease systemic lupus erythematosus, or SLE, because it turns out that lupus patients have too much prolactin, and prolactin makes immune cells active. These new avenues of treatment could not have been devised before the studies were in place that showed that hormones such as prolactin or nerve chemicals such as adrenaline change immune cell function, and thus perhaps immune disease.

Novel as these new treatments are, they are still based on a way of thinking about disease in terms of single chemicals, single cell types, or single organs. But at every moment in time, thousands and millions of chemicals and cells interact and change simultaneously. It is daunting to try to imagine how one might sort out the many reverberating loops of all these cells and chemicals working with one another. The permutations and combinations of possibilities seem infinite. And yet this is the challenge of a field whose goal it is to understand the interactions of the many cells and chemicals in two very complex and tightly interwoven systems. Until now, analyzing one or a few compounds at a time was the best that our technological tools allowed us to do. So scientists had to close their eyes to all the surrounding cells and chemicals, and focus in more and more minute steps on the one they were studying.

New technologies now allow us to measure changes in the products of thousands and thousands of genes at a time, all on a single microchip. This new technology, grown from a marriage of Silicon Valley with molecular biology, builds each spot of DNA directly on a glass microscope slide, using the same principles that computer manufacturers use to build microchips. Extracts of complex biological tissues can then be overlaid on top of the thousands of spots of DNA arrayed on the slide. With fluorescent or radioactive labels, changes in the products of these thousands of genes can then be viewed simultaneously.

The difference in richness of understanding that this technology permits, compared to studying a single molecule, is akin to the difference between looking at the musical score for a symphony, and playing all the notes on the page, or choosing a single note—the same throughout the piece—and playing it alone. Rhythm and pitch are important, but alone they let us hear only one dimension of the piece. It is only by hearing all the notes together, and hearing them change against each other, that the real sound of the music emerges. For it might be that with such new technologies we will discover that in some biological responses and some illnesses, the relationship of all the notes to each other might stay the same while the key in which they are played changes.

We can only imagine what we will discover when such technologies are used to compare the complex patterns of thousands of new gene products that change in the brain's stress center of those rats with a low or a high stress response. Or if the same technology were applied to see the thousands of gene products that are made by immune cells in a drop of blood from patients in the first hours after exposure to infection. In clinical settings, the ability to continuously collect minute quantities of blood, combined with new methods of complex mathematics, allows analysis of the thousands of pulses of hormones that drive the rhythms of our bodies. Such levels of analysis help us to understand illnesses that might involve perturbations not of amounts or forms of hormones, but of the rhythms of their pulse. New brain imaging technologies show not only abnormalities of anatomy, and increases or decreases of activity in different parts of brain, but also the temporal and spatial relationships of the activity in these parts of brain. We can now see which brain centers light up first—emotional, visual, memory, and back and forth again—after the eye perceives an object or after the individual thinks a stressful thought. And the sequence and order and reverberating loops of these brain activities tell us more about the way the brain functions in health and in disease than any single static picture. So, another layer of understanding is added by this knowledge. At the same time, advances in immunology are also helping us measure the sequence over time of the changes of the many different molecules and cells turned on by different kinds of immune triggers. By blending these two sciences—neurobiology and immunology—we will be able to understand, at an increasingly detailed level, how perceptions of events that are colored by emotions can influence different components of the immune response, and ultimately disease. And once we understand that, we can begin to design ways in which modifying our emotional responses might help to prevent or change the course of immune, inflammatory, and infectious diseases.

Understanding the brain–immune connections at these minute and systems levels will help physicians believe their patients when they say that believing and laughter make them well. Because suddenly an ephemeral, ghostlike concept such as belief, which previously could not be grasped in concrete terms, has a chance of being broken down into the many cells and nerve pathways and chemicals that make it happen. This level of knowledge could help physicians understand the ways in which hypnotism, meditation, group therapy, talismans such as crystals, or soothing rituals such as aromatherapy and spas can all help healing through belief. It might even help us understand some ways that prayer can heal.

Such knowledge will help not only physicians and their patients but people in their daily lives. By understanding the biological basis of marital stress and its effect on illness, we can learn better coping strategies that may prevent such situations and their deleterious effects. By understanding the effects of sleep and sleeplessness, nutrition and malnutrition on immune function, we can sort out the parts of stress effects that come from imbalances in these physiological states. And by understanding the biological basis of workplace stress and its effect on illness, perhaps we can design less stressful work environments and more sensitive management techniques that still optimize productivity while maximizing the workers' sense of well-being.

When management at the Volvo factory in Sweden found a high incidence of heart disease, stroke, hypertension, and dissatisfaction in their workers on the assembly line, they incorporated environmental modifications into the workplace to reduce stress. They made the surroundings more pleasant, with open airy spaces, plants, and areas in which workers could relax and socialize. They did away with the boring, repetitive assembly line, replacing it with work groups whose goal was to produce a car. When workers had the finished product as their goal, when they felt valued and respected, and in the new and worker-friendly environment, the incidence of these maladies decreased. At the same time, worker satisfaction and product quality increased. Everyone won.

But not all stress responses and diseases can be done away with solely by changing the environment. Part of the future in this field includes identifying which components of our physiological responses we can modify and which we can't; which can be changed by learning new ways of coping with stressful situations and which are unmodifiable except by medical intervention. Another important future task of this field will be to identify which of our behavioral and physiological responses to stressful situations are inherited, which are learned, and how much is affected by early upbringing. From all these pieces of the puzzle, we will be able to determine which of the effects of the emotions and our stress response on health can be changed, and by how much, with learning-based interventions. By understanding the active components of these soothing practices, physicians will be better able to help people choose approaches that might work best for them and so include them in the overall health management armamentarium.

This science will also help patients who try these routes and fail realize that it is not their fault that they are sick. It could be that in their case, no matter how hard they might try, their genes and preset immune and hormone responses prevent them from relearning modes of responding to stress enough to change their biological responses. In such cases, this science can teach us how to combine medicines derived from understanding the molecules of these connections with practices derived from centuries of use, bolstered now by understanding their scientific underpinnings.

This may all sound like utopia, and perhaps it is too much to expect from such a science. But if this new science accomplishes one single thing, certainly it will be this—it will force us to learn to listen and to speak to each other in each other's languages. It will force scientists and physicians to learn the languages of their colleagues' disciplines. It will help medical specialists think of the person as a whole —one single body with one single soul. It will help physicians speak the language of their patients, and listen to them. And in that learning and that listening we will have crossed the hardest divide of all, the one between the popular culture and science. If we can bridge that gap, then scientists and physicians will accept that popular beliefs have their place and their reasons for being. Through this acceptance, scientists can begin to try to understand the workings of such beliefs, and the public, while still embracing them, can nonetheless begin to understand their limitations. In this meeting place, perhaps this emerging science can help popular culture and medicine meld these two approaches to health and healing. By focusing on the minute connections between each part, and at the same time looking outward to the emotions and beyond, scientists and physicians will be able to work out in intricate detail how the cogs and wheels fit together in an interlocking, time-defined pattern, which, when functioning harmoniously, is the basis of "feeling well."

Bibliography

Ader, R., and Cohen, N. "Behaviorally conditioned immunosuppression and murine systemic lupus erythematosus." *Science* **215** (4539) March 19,1982: 1534–1536.

Ader, R., Cohen, N., and Bovbjerg, D. "Conditioned suppression of humoral immunity in the rat." *J Comp Physiol Psychol* **96** (3) June 1982: 517–521.

Ader, R., Felten, D. L., and Cohen, N., eds. *Psychoneuroimmunology,* 2nd ed. New York: Academic Press, 1999.

Akita, S., Webster, J., Ren, S., Takino, H., Said, J., Zand, O., and Melmed, S. "Human and murine pituitary expression of leukemia inhibitory factor (LIF): Novel intrapituitary regulation of adrenocorticotrophin (ACTH) synthesis and secretion." *J Clin Invest* **95** (1995): 1288–1298.

Antoni M., Baggett, L., Ironson, G., LaPerriere, A., August, S., Klimas, N., Schneiderman, N., and Fletcher, M. "Cognitive-behavioral stress management intervention buffers distress responses and immunologic changes following notification of HIV-1 seropositivity." *J Consult Clin Psychol* **59** (6) (December 1991): 906–915.

Arimura, A., and Takaki, A. "Interactions between cytokines and the hypothalamic–pituitary–adrenal axis during stress." *Ann NY Acad Sci* **739** (270) (1994): 270–281.

Arzt, E., Stelzer, G., Renner, U., Lange, M., Muller, O., and Stalla, K. "Interleukin-2 and IL-2 receptor expression in human corticotrophic adenoma and murine pituitary cell cultures." *J Clin Invest* **90** (1992): 1944–1951.

Axelrod, J., and Reisine, T. "Stress hormones: Their interaction and regulation." *Science* **224** (1984): 452.

Bacon, F. "The New Atlantis." Originally published 1627. Reprinted in The Harvard Classics, vol. 3, ed. Charles W. Eliot, LLD. New York: P. F. Collier and Son, 1909.

Bailey C., Bartsch, D., and Kandel, E. "Toward a molecular definition of long-term memory storage." *Proc Natl Acad Sci USA* **93** (24) (November 26, 1996): 13445–13452.

Balick, M. J., and Cox, P. A. *Plants, People, and Culture: The Science of EthnoBotany.* New York: W. H. Freeman and Company, 1997.

Banks, W. A. "Comparison of saturable transport and extracellular pathways in the passage of interleukin-1 alpha across the blood-brain barrier." *J Neuroimmunol* **67** (1) (1996): 41–47.

Bellinger, D., Lorton, D., Felten, S., and Felten, D. "Innervation of lymphoid organs and implications in development, aging and autoimmunity." *Internatl J Immunopharmacol* **14** (1992): 329–344.

Benson, H. "The relaxation response: Therapeutic effect [letter; comment]." *Science* **278** (5344) (December 1997): 1694–1695.

Benveniste, E. "Cytokine circuits in the brain. Implications for AIDS dementia complex." *Res Publ ANRMD* **72** (1994): 71–88.

Berczi, I., and Szelenyi, J. *Advances in Psychoneuroimmunology,* Vol III. New York: Plenum Publishing Corporation, 1994.

Berkenbosch F., van Oers, J., del Rey, A., Tilders, F., and Besedovsky, H. "Corticotropin-releasing factor-producing neurons in the rat activated by interleukin-1." *Science* **238** (October 23, 1987): 524–526.

Bernton E., Beach, J., Holaday J., Smallridge R., and Fein H. "Release of multiple hormones by a direct action of interleukin-1 on pituitary cells." *Science* **238** (4826) (October 23, 1987): 519–521.

Besedovsky H., del Rey A., and Sorkin E. "Immune-neuroendocrine interactions." *J Immunol* **135** (2 Suppl) (August 1985): 750s–754.

Besedovsky H., del Rey A., Sorkin E., and Dinarello C. "Immunoregulatory feedback between interleukin-1 and glucocorticoid hormones." *Science* **233** (4764) (August 8 1986): 652–654.

Besedovsky, H., Sorkin, E., Keller, M., and Muller, J. "Changes in blood hormone levels during the immune response." *P Soc Exper Bio Med* **150** (1975): 466–467.

Blatteis, C., and Sehic, E. "Prostoglandin E2: A putative fever mediator." In P. Mackowiak (ed.), *Fever: Basic Mechanisms and Management,* 2nd ed. (pp. 117–145). Philadelphia, PA: Lippincott-Raven, 1997.

Blum, A., Elliott, D., Metwali, A., Li, J., Qadir, K., and Weinstock, J. "Substance P regulates somatostatin expression in inflammation." *J Immunol* **161** (11) (December 1, 1998): 6316–6322.

Bluthe, R., Michaud, B., Kelley, K., and Dantzer R. "Vagotomy blocks behavioural effects of interleukin-1 injected via the intraperitoneal route but not via other systemic routes." *Neuroreport* **7** (15–17) (November 4, 1996): 2823–2827.

Boorstin, D. J. *The Creators: A History of Heroes of the Imagination.* New York: Vintage Books, Random House, 1993.

Breder, C., Dinarello, C., and Saper, C. "Interleukin-1 immunoreactive innervation of the human hypothalamus." *Science* **240** (4850) (April 15, 1988): 321–324.

Brenneman, D., Schultzberg, M., Bartfai, T., and Gozes, I. "Cytokine regulation of neuronal survival." *J Neurochem* **58** (2) (February 1992): 454–460.

Brown, D., Lafuse, W., and Zwilling, B. "Host resistance to mycobacteria is compromised by activation of the hypothalamic-pituitary-adrenal axis." *Ann NY Acad Sci* **840** (1998): 773–786.

Brown, T. M. "Descartes, dualism, and psychosomatic medicine." In W. F. Bynum, R. Porter, and M. Shepherd (eds.), *The Anatomy of Madness: Essays in the History of Psychiatry*, Vol. 1, pp. 40–62. London: Tavistock, 1985.

Brown, T. M. "Emotions and Disease in Historical Perspective" in *Emotions and Disease: An Exhibition at the National Library of Medicine.* Exhibition Directors: E. Fee, E. M. Sternberg. Curators: A. Harrington, T. M. Brown. Bethesda, MD. Friends of the National Library of Medicine, 1997.

Bulloch, K., Hausman, J., Radojcic, T., and Short, S. "Calcitonin gene-related peptide in the developing and aging thymus. An immunocytochemical study." *Ann NY Acad Sci* **621** (1991): 218–228.

Burckardt, J. *The Greeks and Greek Civilization.* New York: St. Martin's Press, 1998.

Buske-Kirschbaum, A., Jobst, S., Psych, D., Wustmans, A., Kirschbaum, C., Rauh, W., and Hellhammer, D. "Attenuated free cortisol response to psychosocial stress in children with atopic dermatitis." *Psychosom Med* **59** (4) (July–August, 1997): 419–426.

Cacioppo, J., Berntson, G., Malarkey, W., Kiecolt-Glaser, J., Sheridan, J., Poehlmann, K., Burleson, M., Ernst, J., Hawkley, L., and Glaser, R. "Autonomic, neuroendocrine, and immune responses to psychological stress: The reactivity hypothesis." *Ann NY Acad Sci* **840** (1998): 664–673.

Cacioppo, J. T., Tassinary, L. G., and Berntson, G. G. *Handbook of psychophysiology,* 2nd ed. New York: Cambridge University Press, 2000.

Campbell, I. "Structural and Functional Impact of the Transgenic Expression of Cytokines in the CNS." *Ann NY Acad Sci* **840** (1998): 83–96.

Carter, S. C., Lederhendler, I. I., and Kirkpatrick, B. (eds). *The Integrative Neurobiology of Affiliation:* Annals of the New York Academy of Sciences, Vol. 807. New York: New York Academy of Sciences, 1997.

Catania, A., and Lipton, J. "Alpha-melanocyte stimulating hormone in the modulation of host reactions." *Endocr Rev* **14** (1993): 564–576.

Chrousos, G. "The hypothalamic–pituitary–adrenal axis and immune mediated inflammation." *N Eng J Med* **332** (1995): 1351–1362.

Chrousos, G. P., and Gold, P. W. "The concepts of stress and stress system disorders. Overview of physical and behavioral homeostasis." *JAMA* **267** (9) (March 1992): 1244–1252.

Chuluyan, H., Saphier, D., Rohn, W., and Dunn, A. "Noradrenergic innervation of the hypothalamus participates in adrenocortical responses to interleukin-1." *Neuroendocrinology* **56** (1) (July 1992): 106–111.

Cohen, S., Doyle, W. J., Skoner, D. P., Rabin B. S., and Gwaltney, J. M. "Social ties and susceptibility to the common cold." *JAMA* **277** (24) (June 25, 1997): 1940–1944.

Cohen, S., Tyrrell, D. A., and Smith, A. P. "Psychological stress and susceptibility to the common cold." *N Engl J Med* **325** (9) (August 1991): 606–612.

Conte, H., and Plutchik, R. "A circumplex model for interpersonal personality traits." *J Pers Soc Psych* **40** (4) (1981):701–711.

Conti, A., and Maestroni, G. "Melatonin rhythms in mice: Role in autoimmune and lymphoproliferative diseases." *Ann NY Acad Sci* **840** (1998): 395–410.

Costa, A., Trainer, P., Besser, M., and Grossman, A. "Nitric oxide modulates the release of corticotropin releasing hormone from the rat hypothalamus in vitro." *Brain Res* **605** (1993): 187–192.

Courtney, S. M., Petit, L., Haxby, J. V., and Ungerleider, L. G. "The role of prefrontal cortex in working memory: examining the contents of consciousness." *Philos Trans R Soc Lond B Biol Sci* **353** (1377) November 1998: 1819–1828.

Courtney, S. M., Petit, L., Maisog, J. M., Ungerleider, L. G., and Haxby, J. V. "An area specialized for spatial working memory in human frontal cortex." *Science* **279** (5355) February 27, 1998: 1347–1351.

Crofford, L., Pillemer, S., Kalogeras, K., Cash, J., Michelson, D., Kling, M., Sternberg, E., Gold, P., Chrousos, G., and Wilder, R. "Hypothalamic–pituitary–adrenal axis perturbations in patients with fibromyalgia." *Arthritis Rheum* **37** (1994): 1583–1592.

Cupps, T. R., and Fauci, A.S. "Corticosteroid-mediated immunoregulation in man." *Immunol Rev* **65** (1982): 133–155.

Damasio, A. *Descartes' Error: Emotion, Reason and the Human Brain.* New York: Putnam Publishing, 1994.

Darwin, C. *The Expression of the Emotions in Man and Animals.* London: John Murray, Albermarle Street, 1872.

De Kruif, P. "Leeuwenhook: First of the Microbe Hunters." From *Microbe Hunters.* New York: Harcourt, Brace & Co. Inc, 1926.

del Rey, A., Besedovsky, H., and Sorkin, E. "Endogenous blood levels of corticosterone control the immunologic cell mass and B cell activity in mice." *J Immunol* **133** (2) (August 1984): 572–575.

Demitrack, M., Dale, J., Straus, S., Laue, L., Listwak, S., Kruesi, M., Chrousos, G., and Gold, P. "Evidence for impaired activation of the hypothalamic–pituitary–adrenal axis in patients with chronic fatigue syndrome." *J Clin Endocrinol Metab* **73** (6) (1991): 1224–1234.

Denburg, J. A., Befus, A. D., and Bienenstock, J. "Growth and differentiation in vitro of mast cells from mesenteric lymph nodes of Nippostrongylus brasiliensis-infected rats." *Immunology* **41** (1) (September 1980): 195–202.

Desimone, R. "Neural mechanisms for visual memory and their role in attention." *Proc Natl Acad Sci U S A* **93** (24) November 26, 1996: 13494–13499.

Dhabhar, F., Miller, A., McEwen, B., and Spencer, R. "Stress-induced changes in blood leukocyte distribution–role of adrenal steroid hormones." *J Immunol* **157** (1996): 1638–1644.

Edelstein, E. J., and Edelstein, L. *Asclepius: Collection and Interpretation of the Testimonies.* Baltimore: The Johns Hopkins University Press, 1998.

Ekman, P. "Facial expression of emotion." *American Psychologist* **48** (1993): 384–392.

Ekman, P., and Davidson R. *The Nature of Emotion: Fundamental Questions.* New York: Oxford University Press, 1994.

Engel, G. L., and Reichsman, F. "Spontaneous and induced depression in an infant with gastric fistula." *Journal of the American PsychoAnalytic Association* **4** (1956): 428–452.

Engel, G. L., Reichsman, F., and Segal, H. L. "A study of an infant with a gastric fistula." *Psychosomatic Medicine* **18** (1956): 374–398.

Ericsson, A., Kovacs, K., and Sawchenko, P. "A functional anatomical analysis of central pathways subserving the effects of interleukin-1 on stress-related neuroendocrine neurons." *J Neurosci* **14** (2) (February 1994): 897–913.

Everson, R. M., Prashanth, A. K., Gabbay, M., Knight, B. W., Sirovich, L, and Kaplan, E. "Representation of spatial frequency and orientation in the visual cortex." *Proc Natl Acad Sci U S A* **95** (14) July 7, 1998: 8334–8338.

Fabris, N., Mocchegiani, E., and Provinciali, M. "Pituitary–thyroid axis and immune system: A reciprocal neuroendocrine–immune interaction." *Horm Res* **43** (1–3) (1995): 29–38.

Farrar, W. L., Kilian, P. L., Ruff, M. R., Hill, J. M., and Pert, C. B. "Visualization and characterization of interleukin-1 receptors in brain. *J Immunol* **139** (2) (July 15, 1987): 459–463.

Fawzy, F., Fawzy, N. W., Hyun, C. S., Elashoff, R., Guthrie, D., Fahey, J. L., and Morton, D. "Malignant melanoma. Effects of an early structured psychiatric intervention, coping, and affective state on recurrence and survival 6 years later." *Arch Gen Psychiatry* **50** (9) (September 1993): 681–689.

Fendt, M., and Fanselow, M. S. "The neuroanatomical and neurochemical basis of conditioned fear." *Neurosci Biobehav Rev* **23** (5) May 1999: 743–760.

Fleshner, M., Goehler, L., Hermann, J., Relton, J. K., Maier, S., and Watkins, L. "Interleukin-1 beta induced corticosterone elevation and hypothalamic NE depletion is vagally mediated." *Brain Res Bull* **37** (6) (1995): 605–610.

Fontana, A., Weber, E., and Dayer, J. M. "Synthesis of interleukin 1/endogenous pyrogen in the brain of endotoxin-treated mice: A step in fever induction?" *J Immunol* **133** (4) (October 1984): 1696–1698.

Francis, D., Diorio, J., Liu, D., and Meaney, M. "Nongenomic transmission across generations of maternal behavior and stress responses in the rat." *Science* **286** (November 5, 1999): 1155–1158.

Fukata, J., and Usui, T. "Effects of recombinant human interleukin-1a, -1b, 2 and 6 on ACTH synthesis and release in the mouse pituitary tumour cell line ATT-20." *J Endocrinol* **122** (1988): 33–39.

Funder, J. "Mineralocorticoids, glucocorticoids, receptors and response elements." *Science* **259** (1993): 1132–1133.

Gaillard, R. "Neuroendocrine-immune system interaction. The immune–hypothalamo–pituitary–adrenal axis." *Trends Endocrinol Metab (TEM)* **5** (7) (1994): 303–309.

Gallagher, M., Landfield, P. W., McEwen, B., Meaney, M. J., Rapp, P. R., Sapolsky, R., and West, M. J. "Hippocampal neurodegeneration in aging [letter; comment]." *Science* **274** (5287) (October 25, 1996): 484–485.

Geenen, V., Martens, H., Vandersmissen, E., Achour, I., Kecha, O., and Franchimont, D. "Cellular and molecular aspects of thymic T-cell education in neuroendocrine self principles: Implications for autoimmunity." *Ann NY Acad Sci* **840** (1998): 328–337.

Glaser, R., and Kiecolt-Glaser, J. *Handbook of Stress and Immunity*. San Diego: Academic Press, 1994.

Green, P. G., Janig, W., and Levine, J. D. "Negative feedback neuroendocrine control of inflammatory response in the rat is dependent on the sympathetic postganglionic neuron." *J Neurosci* **17** (9) (May 1,1997): 3234–3238.

Griffin, A., and Whitacre, C. "Sex and strain differences in the circadian rhythm fluctuation of endocrine and immune function in the rat: Implications for rodent models of autoimmune disease." *J Neuroimmunol* **35** (1991): 53–64.

Griffin D. "Cytokines in the brain during viral infection: clues to HIV-associated dementia." *J Clin Invest* **100** (12) (December 15, 1997): 2948–2951.

Grimm, M., Ben-Baruch, A., Taub, D., Howard, O., Wang, J., and Oppenheim, J. "Opiate inhibition of chemokine-induced chemotaxis." *Ann NY Acad Sci* **840** (1998): 9–20.

Gunnar, M. "Early adversity and the development of stress reactivity regulation." In C. A. Nelson (ed.), *The Effects of Adversity on Neurobehavioral Development* (Minnesota Symposia on Child Psychology, Vol. 31). NJ: Lawrence Erlbaum, 1998.

Hadden, J. "Thymic endocrinology and prospects for treating thymic involution." *Immunopharmacology Reviews II* **2** (1995): 353–370.

Harbuz, M., Rees, R., Echland, D., Jessop, D., Brewerton, D., and Lightman, S. "Paradoxical responses of hypothalamic corticotropin-releasing factor (CRF) messenger ribonucleic acid (mRNA) and CRF-41 peptide and adenohypophysial proopiomelanocortin mRNA during chronic inflammatory stress." *Endocrinology* **130** (1992): 1394–1400.

Harbuz, M., Rees, R., and Lightman, S. "HPA axis responses to acute stress and adrenalectomy during adjuvant-induced arthritis in the rat." *Am J Physiol* **264** (1 Pt 2) (1993): r179–185.

Harrington, A. *The Placebo Effect: An Interdisciplinary Exploration.* Cambridge, MA: Harvard University Press, 1999.

Hobson, J. A. *Sleep.* New York: W. H. Freeman and Company, 1995.

Hori, T., Oka, T., Hosoi, M., and Aou, S. "Pain modulatory actions of cytokines and prostaglandin E2 in the brain." *Ann NY Acad Sci* **840** (1998): 269–281.

Hyman, S. E. "Brain neurocircuitry of anxiety and fear: implications for clinical research and practice." *Biol Psychiatry* **44** (12) December 15, 1998: 1201–1203.

Imaki, T., Nahan, J., Rivier, C., Sawchenko, P., and Vale, W. "Differential regulation of corticotropin-releasing factor mRNA in rat brain regions by glucocorticoids and stress." *J Neurosci* **11** (3) (1991): 585–599.

Irwin, M. "Immune correlates of depression." *Adv Exp Med Biol* **461** (1999): 1–24.

Jafarian-Tehrani, M., Gabellec, M., Adyel, F., Simon, D., Griffais, R., Ternynck, T., and Haour, F. "Interleukin-1 receptor defiency in the hippocampal formation of (NZBxNZW)F2 mice: Genetic and molecular studies relating to autoimmunity." *J Neuroimmunol* **84** (1998): 30–39.

Jonakait, G. "Neural-immune interactions in sympathetic ganglia." *Trends Neurosci* **16** (10) (October 1993): 419–423.

Karalis, K., Sano, H., Redwine, J., Listwak, S., Wilder, R., and Chrousos, G. "Autocrine or paracrine inflammatory actions of corticotropin-releasing hormone in vivo." *Science* **254** (1991): 421–423.

Kent, S., Bluthe, R. M., Kelley, K. W., and Dantzer, R. "Sickness behavior as a new target for drug development." *Trends Pharmacol Sci* **13** (1) (January 1992): 24–28.

Kiecolt-Glaser, J., Fisher, L., Ogrocki, P., Stout, J., Speicher., C., and Glaser, R. "Marital quality, marital disruption, and immune function." *Psychosom Med* **49** (1) (1987): 13–34.

Kiecolt-Glaser, J. K., Glaser, R., Gravenstein, S., Malarkey, W. B., and Sheridan, J. "Chronic stress alters the immune response to influenza virus vaccine in older adults." *Proc Natl Acad Sci USA* **93** (7) (April 1996): 3043–3047.

Kiecolt-Glaser, J., Glaser, R., Shuttleworth, E., Dyer, C., Ogrocki, P., and Speicher, C. "Chronic stress and immunity in family caregivers of Alzheimer's disease victims." *Psychosom Med* **49** (5) (September–October 1987): 523–535.

King, L., Vacchio, S., Hunziker, R., Margulies, D., and Ashwell, J. "A targeted glucocorticoid receptor antisense transgene increases thymocyte apoptosis and altered thymocyte development." *Immunity* **3** (1995): 647–656.

Kirschbaum, C., Strasburger, C. J., Jammers, W., and Hellhammer, D. H. "Cortisol and behavior: 1. Adaptation of a radioimmunoassay kit for reliable and inexpensive salivary cortisol determination." *Pharmacol Biochem Behav* **34** (4) (December 1989): 747–751.

Koenig, H. G., Cohen, H. J., George, L. K., Hays, J. C., Larson, D. B., and Blazer, D. G. "Attendance at religious services, interleukin-6, and other biological parameters of immune function in older adults." *Int J Psychiatry Med* **27** (3) (1997): 233–250.

Kosslyn, S. M. "Aspects of a cognitive neuroscience of mental imagery." *Science* **240** (4859) (June 17, 1988): 1621–1626.

Kostoglou-Athanassiou, R., Jacobs, R., Satta, M., Dahia, P., Costa, A., Navarra, P., Chew, S., Forsling M., and Grossman, A. "Acute and subacute effects of endotoxin on hypothalamic gaseous neuromodulators." *Ann NY Acad Sci* **840** (1998): 240–261.

Kroemer, G., Brezinschek, H. P., Faessler, R., Schauenstein, K., and Wick, G. "Physiology and pathology of an immunoendocrine feedback loop." *Immunol Today* **9** (6) (June 1988): 163–165.

Krueger, J. "Microbial products and cytokines in sleep and fever regulation." *Crit Rev Immunol* **14** (1994): 355–379.

Kuis, W., DeJong-De Vos Van Steenwijk, C., Sinnema, G., Kavelaars, A., Prakken, B., Helders, P., and Heijnen, C. "The autonomic nervous system and the immune system in juvenile chronic rheumatoid arthritis: An interdisciplinary study." *Brain Behav Immun* **10** (4) (1996): 387–398.

Lang, P. J., Kozak, M. J., Miller, G. A., Levin, D. N., and McLean, A., Jr. "Emotional imagery: Conceptual structure and pattern of somato-visceral response." *Psychophysiology* **2** (March 17, 1980): 179–192.

LeDoux, J. "Emotion, memory and the brain." *Sci Am* **270** (6) (June 1994): 50–57.

LeDoux, J. *The Emotional Brain: The Mysterious Underpinnings of Emotional Life.* New York: Simon and Shuster, 1996.

Lesnikov, V., Korneva, A., Dall'Ara, A., and Pierpaoli, W. "The involvement of the pineal gland and melatonin in immunity and aging. II. Thyrotropin releasing hormone and melatonin forestall involution and promote reconstitution of the thymus in anterior hypothalamic area (AHA) lesion mice." *Int J Neurosci* **62** (1992): 141–153.

Levine, J., Dardick, S., Roizen, M., Helms, C., and Basbaum, A. "Contribution of sensory afferents and sympathetic efferents to joint injury in experimental arthritis." *J Neurosci* **6** (1986): 3423–3429.

Licinio, J., and Wong, M. "Pathways and mechanisms for cytokine signaling of the central nervous system." *J Clin Invest* **100** (12) (December 15, 1997): 2941–2947.

Lindgren, K. A., Larson, C. L., Schaefer, S. M., Abercrombie, H. C., Ward, R. T., Oakes T. R., Holden, J. E., Perlman, S. B., Benca, R. M., and Davidson, R. J. "Thalamic metabolic rate predicts EEG alpha power in healthy control subjects but not in depressed patients." *Biol Psychiatry* **45** (8) (April 15, 1999): 943–952.

Linthorst, A., Flachskamm, C., Muller-Pruss, P., Holsboer, F., and Reul, J. "Effect of bacterial endotoxin and interleukin-1B on hippocampal serotonergic neurotransmission, behavioral activity, and free corticosterone levels: An in-vivo microdialysis study." *J Neurocsi* **15** (1995): 2920–2934.

Lovallo, W. R. *Stress and Health: Biological and Psychological Interactions.* Thousand Oaks, CA: SAGE Publications, 1997.

Madden, K., Felten, S., Felten, D., Hardy, C., and Livnat, S. "Sympathetic nervous system modulation of the immune system. II. Induction of lymphocyte proliferation and migration in vivo by chemical sympathectomy." *J Neuroimmunol* **49** (1994): 67–75.

Madden, K., Moynihan, J., Brenner, G., Felten, S., Felten, D., and Livnat, S. "Sympathetic nervous system modulation of the immune system. III. Alterations in T and B cell proliferation and differentiation in vitro following chemical sympathectomy." *J Neuroimmunol* **49** (1994): 77–87.

Madden, K. S., Bellinger, D. L., Felten, S. Y., Snyder, E., Maida, M. E., and Felten, D. L. "Alterations in sympathetic innervation of thymus and spleen in aged mice." *Mech Ageing Dev* **94** (1–3) (March 1997): 165–175.

Marchetti, B., Gallo, F., Farinella, Z., Tarolo, C., Testa, N., Romeo, C., and Morale, M. "Luteinizing hormone-releasing hormone is a primary signaling molecule in the neuroimmune network." *Ann NY Acad Sci* **840** (1998): 205–248.

Mason, D., MacPhee, I., and Antoni F. "The role of the neuroendocrine system in determining genetic susceptibility to experimental allergic encephalomyelitis in the rat." *Immunology* **70** (1990): 1–5.

Mayer, E. A., and Saper, C. B. *The Biological Basis for Mind-Body Interactions.* New York: Elsevier Press, 1999.

McCann, S., Sternberg, E., Lipton, J., Chrousos, G., Gold, P., and Smith, C. (eds.). *Neuroimmunomodulation.* Annals of the New York Academy of Sciences, Vol. 840. New York: New York Academy of Sciences, 1998.

McClintock, M. "Menstrual synchrony and suppression." *Nature* **229** (5282) (January 22, 1971): 244–245.

McEwen, B. "Protective and damaging effects of stress mediators." *N Engl J Med* **338** (3) (1998): 171–179.

Melzack, R. "Sensory modulation of pain." *Int Rehabil* **1** (3) (1979): 111–115.

Michelson, D., Stone, L., Galliven, E., Magiakou, M., Chrousos G., Sternberg, E., and Gold, P. "Multiple sclerosis is associated with alterations in hypothalamic–pituitary–adrenal axis function." *J Clin Endocrinol Metab* **79** (1994): 848–853.

Michelson, D., Stratakis, C., Hill, L., Reynolds, J., Galliven, E., Chrousos, G., and Gold, P. "Bone mineral density in women with depression." *N Engl J Med* **335** (1996): 1176–1181.

Mishkin, M., and Murray, E. A. "Stimulus recognition." *Curr Opin Neurobiol* **4** (2) April 1994: 200–206.

Moalem, G., Leibowitz-Amit, R., Yoles, E., Mor, F., Cohen, I., and Schwartz, M. "Autoimmune T cells protect neurons from secondary degeneration after central nervous system axotomy." *Nat Med* **5** (1) (January 1999): 49–55.

Moisan, M., and Courvoisier, H. "A major quantitative trait locus influences hyperactivity in the WKHA rat." *Nat Genet* **14** (4) (1996): 471–473.

Mucke, L., Masliah, E., and Campbell, I. "Transgenic models to assess the neuropathogenic potential of HIV-1 proteins and cytokines." *Curr Top Microbiol Immunol* **202** (187) (1995): 187–205.

Munck, A., Guyre, P., and Holbrook N. "Physiological functions of glucocorticoids in stress and their relation to pharmacological actions." *Endocr Rev* **5** (1984): 25–44.

Neveu, P. "Brain lateralization and immunomodulation." *Int J Neurosci* **70** (1–2) (May 1993): 135–143.

Ottaway, C., and Stanisz, A. "Neural-immune interactions in the intestine: Implications for inflammatory bowel disease." In J. B. Kirsner and

R. G. Shorter (eds.), *Inflammatory Bowel Disease* (pp. 281–300). Baltimore, MD: Williams and Wilkins, 1995.

Padgett, D., Sheridan, J., Dorne, J., Berntson, G., Candelora, J., and Glaser, R. "Social stress and the reactivation of latent herpes simplex virus type 1." *Proc Natl Acad Sci USA* **95** (12) (June 9, 1998): 7231–7235.

Payan, D. G., and Goetzl, E. J. "Dual roles of substance P: Modulator of immune and neuroendocrine functions." *Ann NY Acad Sci* **512** (1987): 465–475.

Pert, C., and Chopra, D. *Molecules of Emotion: Why You Feel the Way You Feel.* New York: Simon & Schuster, 1999.

Pert, C. B., Kuhar, M. J., and Snyder, S. H. "Opiate receptor: Autoradiographic localization in rat brain." *Proc Natl Acad Sci* **73** (10) (October 1976): 3729–3733.

Petersen, S. E., Fox, P. T., Snyder, A. Z., and Raichle, M. E. "Activation of extrastriate and frontal cortical areas by visual words and word-like stimuli." *Science* **249** (4972) (August 31): 1041–1044.

Plotsky, P. M., and Vale, W. "Patterns of growth hormone-releasing factor and somatostatin secretion into the hypophysial-portal circulation of the rat." *Science* **230** (4724) (October 25, 1985): 461–463.

Porter, R., and Teich, M. *Drugs and Narcotics in History.* London: Cambridge University Press, 1995.

Postel-Vinay, M., Mello-Coehlo, V., Gagnerault, M., and Dardenne, M. "Growth hormone stimulates the proliferation of activated mouse T lymphocytes." *Endocrinology* **138** (1997): 1816–1820.

Pothoulakis, C., Castagliuolo, I., and Leeman, S. "Neuroimmune mechanisms of intestinal responses to stress: Role of corticotropin-releasing factor and neurotensin." *Ann NY Acad Sci* **840** (1998): 635–648.

Proust, M. *In Search of Lost Time,* vol. 1, "Swann's Way," translated by C. K. Scott Moncrieff and Terence Kilmartin. New York: The Modern Library, Random House, 1992.

Rabin, B. *Stress, Immune Function, and Health: The Connection.* New York: Wiley-Liss, 1999.

Raichle, M. "Visualizing the mind." *Scientific American* **270** (April 1994): 58–64.

Reed, G. M., Kemeny, M. E., Taylor, S. E., and Visscher, B. R. "Negative HIV-specific expectancies and AIDS-related bereavement as predictors of symptom onset in asymptomatic HIV-positive gay men." *Health Psychol* **18** (4) (July 1999): 354–363.

Reichlin, S. "Endocrine–immune interaction." In L. J. DeGroot (ed.), *Endocrinology* (pp. 2964–3012). Philadelphia, PA: W.B. Saunders, 1995.

Rettori, V., Belova, N., Dees, W., Nyberg, C., Gimeno, M., and McCann, S. "Role of nitric oxide in the control of luteinizing hormone-releasing hormone release in vivo and in vitro." *Proc Natl Acad Sci USA* **90** (1993): 10130–10134.

Rose, N. R. "The discovery of thyroid autoimmunity." *Immunol Today* **2** (51) (May 1991): 167–168.

Rose, R. M., Jenkins, C. D., and Hurst, M. W. "Health change in air traffic controllers: A prospective study. I. Background and description." *Psychosom Med* **40** (2) (March 1978): 142–165.

Rothwell, N., Allan, S., and Toulmond, S. "The role of interleukin 1 in acute neurodegeneration and stroke: Pathophysiological and therapeutic implications." *J Clin Invest* **100** (11) (December 1, 1997): 2648–2652.

Saphier, D. "Neuroendocrine effects of interferon-alpha in the rat." *Adv Exp Med Biol* **373** (209) (1995):209–218.

Sapolsky, R., Krey, L., and McEwen, B. "The neuroendocrinology of stress and aging: The glucocorticoid cascade hypothesis." *Endocr Rev* **7** (1986): 284–301.

Sapolsky, R., Rivier, C., Yamamoto, G., Plotsky, P., and Vale, W. "Interleukin-1 stimulates the secretion of hypothalamic corticotropin-releasing factor." *Science* **238** (4826) (October 23, 1987): 522–524.

Sapolsky, R. M. *Why Zebras Don't Get Ulcers*. New York: W. H. Freeman and Company, 1998.

Savino, W., Villa-Verde, D., Alves., L., and Dardenne, M. "Neuroendocrine control of the thymus." *Ann NY Acad Sci* **840** (1998): 470–479.

Schultz, W. "Dopamine neurons and their role in reward mechanisms." *Curr Opin Neurobiol* **7** (2) April 1997: 191–197.

Segerstrom, S., Solomon, G., Kemeny, M., and Fahey, J. "Relationship of worry to immune sequelae of the Northridge earthquakes." *J Behav Med* **21** (5) (October 1998): 433–450.

Sei, Y., Makino, M., Vitkovic, L., Chattopadhyay, S. K., Hartley, J. W., and Arora, P. K. "Central nervous system infection in a murine retrovirus-induced immunodeficiency syndrome." *J Neuroimmunol* **37** (1992): 131–140.

Selye, H. *The Physiology and Pathology of Exposure to Stress*. Montreal: ACTA, Inc., 1950.

Selye, H. *The Stress of Life*. New York: McGraw-Hill, 1956.

Shapiro, A. K., and Shapiro, E. *The Powerful Placebo*. Baltimore, MD: The Johns Hopkins University Press, 1997.

Silver, R., Silverman, A., Vitkovic, L., and Lederhendler, I. "Mast cells in the brain: Evidence and functional significance." *Trends Neurosci* **19** (1) (January 1996): 25–31.

Sorkin, E. "Changes in blood hormone levels during immune response" *Pro. Soc Exper Bio Med* **150** (1975): 466–467.

Spector, N. "Neuroimmunomodulation: A brief review. Can conditioning of natural killer cell activity reverse cancer and/or aging?" *Regul Toxicol Pharmacol* **24** (1 Pt 2) (August 1996): S32–S38.

Spiegel, D., Bloom, J., Kraemer, H., and Gottheil, E. "Effect of psychosocial treatment on survival of patients with metastatic breast cancer." *Lancet* **2** (8668) (October 14, 1989): 888–891.

Spiegel, K., Leproult, R., and Van Cauter, E. "Impact of sleep debt on metabolic and endocrine function." *Lancet* **354** (9188) (October 23, 1999): 1435–1439.

Stanisz, A. M., Scicchitano, R., Dazin, P., Bienenstock, J., and Payan, D. G. "Distribution of substance P receptors on murine spleen and Peyer's patch T and B cells." *J Immunol* **139** (3) (August 1, 1987): 749–754.

Sternberg, E. "Emotions and disease: From balance of humors to balance of molecules." *Nat Med* **3** (3) (1997): 264–267.

Sternberg, E. "Cytokines and the brain. Neural–immune interactions in health and disease. Perspectives series." *J Clin Invest* **100** (11) (1997): 1–7.

Sternberg, E., Chrousos, G., Wilder, R., and Gold, P. "The stress response and the regulation of inflammatory disease." *Ann Intern Med* **117** (10) (November 15, 1992): 854–866.

Sternberg, E., and Gold, P. "The mind-body interaction in disease." *Scientific American* **7** (1) (1997): 8–15.

Sternberg, E., Hill, J., Chrousos, G., Kamilaris, T., Listwak, S., Gold, P., and Wilder R. "Inflammatory mediator-induced hypothalamic-pituitary-adrenal axis activation is defective in streptococcal cell wall arthritis-susceptible Lewis rats." *Proc Natl Acad Sci USA* **86** (7) (April 1989): 2374–2378.

Sternberg, E., Young, W., Bernardini, R., Calogero, A., Chrousos, G., Gold, P., and Wilder, R. "A central nervous system defect in biosynthesis of corticotropin-releasing hormone is associated with susceptibility to streptococcal cell wall induced arthritis in Lewis rats." *Proc Natl Acad Sci* **86** (12) (June 1989): 4771–4775.

Sundar, S., Becker, K., Cierpial, M., Carpenter, M., Rankin, L., Fleener, S., Ritchie, J., Simson, P., and Weiss, J. "Intracerebroventricular infusion of interleukin 1 rapidly decreases peripheral cellular immune responses." *Proc Natl Acad Sci USA* **86** (16) (1989): 6398–6402.

Swanson, L., Sawchenko, P., Rivier, J., and Vale, W. "Organization of ovine corticotropin-releasing factor immunoreactive cells and fibers in the rat brain: An immunohistochemical study." *Neuroendocrinology* **36** (3) (1983): 165–186.

Tabira, T., Chui, D., Fan, J., Shirabe, T., and Konishi, Y. "Interleukin-3 and interleukin-3 receptors in the brain." *Ann NY Acad Sci* **840** (1998): 107–122.

van Oers, H., de Kloet, E., and Levine, S. "Persistent effects of maternal deprivation on HPA regulation can be reversed by feeding and stroking, but not by dexamethasone." *J Neuroendocrinol* **11** (8) (August 1999): 581–588.

Walker, S., Allan, S., Hoffman, R., and McMurray, R. "Prolactin: a stimulator of disease activity in systemic lupus erythematosus." *Lupus* **4** (1995): 3–9.

Watkins, L., Goehler, L., Relton, J., Tartaglia, N., Silbert, L., Martin, D., and Maier, S. "Blockade of interleukin-1 induced hyperthermia by subdiaphragmatic vagotomy: Evidence for vagal mediation of immune-brain communication." *Neurosci Lett* **183** (1–2) (1995): 27–31.

Watkins, L. R., Maier, S. F., and Goehler, L. "Cytokine-to-brain communication: A review & analysis of alternative mechanisms." *Life Sci* **57** (11) (1995): 1011–1026.

Weiss, J., Sundar, S., Becker, K., and Cierpial, M. "Behavioral and neural influences on cellular immune responses: Effects of stress and interleukin-1." *J Clin Psychiatry* **50** (May 1989): 43–55.

Wiegers, G., Croiset, J., Reul, J., Holsboer, F., and Kloet, E. "Differential effects of corticosteroids on rat peripheral blood T-lymphocyte mitogenesis in vivo and in vitro." *Am J Phys* **155** (December 1993): 1893–1902.

Wilson, M. A., and Tonegawa, S. "Synaptic plasticity, place cells and spatial memory: study with second generation knockouts." *Trends Neurosci* **20** (3) March 1997: 102–106.

Woloski, B. M. R. N. J., Smith, E. M., Meyer, W. J., III, Fuller, G. M., and Blalock, J. E. "Corticotropin-releasing activity of monokines." *Science* **230** 1985:1035–1037.

Yehuda, R. "Stress and glucocorticoid [letter; comment]." *Science* **275** (5306) (March 14, 1997): 1662–1663.

Sources of Illustrations

engraving, Courtesy of the History of Medicine Division, National
Library of Medicine. Bethesda, MD.

Chapter 4

p. 58   William A. Mapes Illustration/Design. Gaithersburg, MD.

p. 65   Photograph with permission, Istvan Berczi, M. D., University of
Manitoba, Department of Immunology, Faculty of Medicine,
Winnepeg, Manitoba, Canada.

Chapter 5

p. 89   William A. Mapes Illustration/Design. Gaithersburg, MD.

Chapter 6

p. 101   Roman bath. Culver Pictures.

Chapter 8

p. 142   From Duchenne de Bologne, *The Mechanism of Human Facial Expression
(Studies in Emotion and Social Interaction),* translated by Andrew R.
Cuthbertson. New York: Cambridge University Press, 1990.

Chapter 9

p. 168   Etching from François Pierre Chaumeton, 1775–1819, in *Pavot.*
Paris: Panckoucke 1815–1820. Courtesy of the History of
Medicine Division, National Library of Medicine. Bethesda, MD.

p. 174   From A. Tosini, Veduta dell'Orto Botanico (da A. Ceni, *Guida
all'Imperial Regio Orto Botanico in Padova,* Padova, 1854). With
permission from Universita degli Studi di Padova, Orto Botanico,
University of Padova, Padova, Italy.

Chapter 11

p. 202   Albert Herter, ca. 1924. Mural, Great Hall, National Academy of
Sciences, Washington, DC. With permission, National Academy of
Sciences. Photographer: Robert C. Lautman, Washington, DC.

Credits

p. 11 "Adelaide's Lament" from *Guys and Dolls* by Frank Loesser (1910–1969), ©1950 Frank Music Corporation, New York, NY. With permission, estate of Frank Loesser.

p. 33 Marcel Proust from *In Search of Lost Time,* vol. 1, "Swann's Way," translated by C. K. Scott Moncrieff and Terence Kilmartin. New York: The Modern Library, 1992 edition published by Random House, Inc.

p. 113 Passage from a letter home by Corporal Richard F. Sutter, USMC, a Marine Infantryman in combat for 13 months, 1966–1967. From a story by Phil McCombs, *Washington Post* staff writer, originally published in the *Washington Post,* July 19, 1998, p. A-17. With permission, *Washington Post,* Washington, DC, and Robert J. Sutter, former Captain, USMC.

p. 164 From Eugene F. Dubois, "The Use of Placebos in Therapy." Cornell Conferences on Therapy, 1946.

pp. 199, 200 Francis Bacon, "The New Atlantis," 1623. First edition published by Dr. Rawley, 1627. From vol. 3, The Harvard Classics, edited by Charles W. Eliot, LLD. New York: P. F. Collier and Son, 1909.

Index